Electrical Engineering for Beginners

From Zero to Pro in 7 Days: Master Core Concepts with Step-by-Step Guidance, Build Functional Projects, and Q&A for Skill Sharpening

Lance Barton

© Copyright 2024 by Lance Barton. All Rights Reserved.

The publication is sold with the idea that the publisher is not required to render accounting, officially permitted or otherwise qualified services. This document is geared towards providing exact and reliable information concerning the topic and issue covered. If advice is necessary, legal or professional, a practiced individual in the profession should be ordered. From a Declaration of Principles which was accepted and approved equally by a Committee of the American Bar Association and a Committee of Publishers and Associations.

In no way is it legal to reproduce, duplicate, or transmit any part of this document in either electronic means or printed format. Recording of this publication is strictly prohibited, and any storage of this document is not allowed unless with written permission from the publisher—all rights reserved.

The information provided herein is stated to be truthful and consistent. Any liability, in terms of inattention or otherwise, by any usage or abuse of any policies, processes, or directions contained within is the sole and utter responsibility of the recipient reader. Under no circumstances will any legal responsibility or blame be held against the publisher for any reparation, damages, or monetary loss due to the information herein, either directly or indirectly. Respective authors own all copyrights not held by the publisher.

The information herein is offered for informational purposes solely and is universal as so. The presentation of the information is without a contract or any guarantee assurance. The trademarks that are used are without any consent, and the publication of the trademark is without permission or backing by the trademark owner. All trademarks and brands within this book are for clarifying purposes only and are owned by the owners themselves, not affiliated with this document

Table of contents

Introduction 1

Importance of Electrical Engineering 1
Who This Book is For 1
Overview of the Book 2

Chapter 1: Basics of Electricity 3

1.1 What is Electricity? 3
1.2 Voltage, Current, and Resistance: The Big Three 4
1.3 Ohm's Law and Basic Calculations 8
1.4 Electric Circuits: Series vs. Parallel 11
1.5 Power and Energy in Circuits 14
1.6 Safety Precautions When Working with Electricity 18

Chapter 2: Understanding Electrical Components. 22

2.1 Resistors, Capacitors, and Inductors 22
2.2 Diodes and Transistors 25
2.3 Integrated Circuits 28
2.4 Switches, Relays, and Connectors 32
2.5 Batteries and Power Supplies 34
2.6 Real-World Examples of Component Usage 38

Chapter 3: Fundamentals of Circuit Design 44

3.1 What Makes a Circuit Work? 44
3.2 Basic Tools for Circuit Design and Testing 47
3.3 Building a Simple Circuit: Step-by-Step 51
3.4 Common Issues and How to Troubleshoot Them 55
3.5 Introduction to Breadboards and Soldering 59

3.6 Tips for Organizing and Documenting Your Work..............63

Chapter 4: Introduction to Electrical Systems 68

4.1 Power Generation and Distribution 68

4.2 AC vs. DC Current: What You Need to Know 71

4.3 Transformers and Their Role in Power Systems 75

4.4 Electric Motors and Generators... 80

4.5 Basic Overview of Renewable Energy Systems 84

4.6 Everyday Applications of Electrical Systems 89

Chapter 5: Introduction to Electronics 92

5.1 The Role of Electronics in Electrical Engineering................92

5.2 Digital vs. Analog Signals..94

5.3 Basics of Logic Gates and Boolean Algebra 98

5.4 Introduction to Microcontrollers and Embedded Systems 102

5.5 Practical Applications of Electronics in Everyday Life 108

Chapter 6: Careers and Further Learning in Electrical Engineering 114

6.1 Typical Career Paths for Electrical Engineers.....................114

6.2 Skills and Certifications to Advance Your Career 117

6.3 Resources for Lifelong Learning (Books, Online Courses, Communities) 121

6.4 Tips for Success in Electrical Engineering......................... 127

Chapter 7: Practical Projects in Electrical Engineering 129

7.1 Why Hands-On Experience is Crucial................................ 129

7.2 Creating a Dynamic LED Lighting System 131

7.3 Building a Solar-Powered USB Charger............................. 136

7.4 Designing a Temperature Sensor with Microcontroller Integration 139

7.5 Constructing a DC Motor Control System 143

7.6 Final Tips for Taking Your Projects Further 148

Chaprt 8: The Future of Electrical Engineering 151

8.1 Emerging Technologies in Electrical Engineering 151

8.2 The Role of Sustainability and Renewable Energy 155

8.3 Challenges Facing the Industry 157

8.4 Opportunities for the Next Generation of Engineers 160

Conclusion 166

Bonus 167

Introduction

Importance of Electrical Engineering

Electrical engineering is a cornerstone of modern life, enabling the technological advancements that shape the world around us. From powering homes and businesses to driving innovation in healthcare, transportation, and communication, it plays a critical role in improving our quality of life. Imagine a world without electricity: no lights, smartphones, or internet. Electrical engineering makes these conveniences possible, and its applications continue to grow with advances in renewable energy, robotics, and artificial intelligence.

This field is more than just wires and circuits; it's about solving real-world problems and creating solutions that make lives better. Electrical engineers design and maintain systems that power industries, ensure safety in critical systems, and drive progress toward a sustainable future. Whether it's designing an electric vehicle or implementing a smart grid, electrical engineering is at the heart of innovation.

By understanding its importance, you'll see why learning the basics is a worthwhile investment. Whether you want to build your first circuit or simply appreciate how your devices work, electrical engineering opens up a world of possibilities.

Who This Book is For

This book is for anyone curious about the world of electrical engineering, especially beginners. You might be a student exploring potential career paths, a hobbyist eager to dive into electronics, or someone with no technical background but a desire to learn something new.

The content assumes no prior knowledge and starts from the very basics. Step-by-step explanations and real-world examples make even complex topics accessible. If you've ever felt intimidated by technical jargon, this book is designed to demystify those concepts and give you the confidence to explore further.

Whether your goal is to build practical skills, gain foundational knowledge, or simply understand how electricity powers your world, this book is your starting point.

Overview of the Book

This book takes you on a journey through the fundamentals of electrical engineering, offering a balance of theory and practical knowledge. We'll start with the basics of electricity, introducing essential concepts like voltage, current, and resistance. Then, we'll explore the components that form the building blocks of electrical systems, such as resistors, capacitors, and transistors.
In later chapters, you'll learn about circuit design, including how to build, test, and troubleshoot simple circuits. We'll also delve into larger electrical systems, like power generation and distribution, and compare AC and DC currents. For those interested in electronics, you'll discover topics like microcontrollers, logic gates, and practical applications of electronics in everyday life.
Finally, we'll discuss career opportunities in electrical engineering, with tips for further learning and success in the field. Along the way, you'll find examples, illustrations, and exercises to reinforce your understanding. By the end of the book, you'll have the foundational knowledge to pursue further studies or dive into hands-on projects.
This book isn't just about learning—it's about discovering how electrical engineering shapes the world and how you can be a part of it. Let's get started!

Chapter 1: Basics of Electricity

1.1 What is Electricity?

Electricity is a form of energy created by the movement of charged particles, primarily electrons. At its core, electricity is the flow of electric charge through a material, enabling the operation of devices, the illumination of homes, and the powering of industries. Understanding electricity starts with examining its behavior at the atomic level.

Atoms, the basic building blocks of matter, consist of a nucleus made up of positively charged protons and neutral neutrons, surrounded by negatively charged electrons. Electrons move in orbits around the nucleus but can sometimes be knocked free, creating an imbalance of charge. When free electrons move through a material, such as a metal conductor, an electric current is produced. This flow of electrons forms the essence of electricity.

There are two main types of electricity: static and current. Static electricity is the accumulation of electric charge on the surface of an object, which can discharge suddenly, as seen when you experience a shock after walking on a carpet. While fascinating, static electricity is less practical for everyday use. The electricity we rely on daily is current electricity, which involves a continuous flow of electrons through a conductor, such as a copper wire.

Current electricity is divided into two forms: direct current (DC) and alternating current (AC). Direct current flows in a single, steady direction, making it ideal for batteries and small electronic devices. Alternating current, however, periodically reverses direction. This type of electricity is used in homes, offices, and industries because it is more efficient for transmitting power over long distances.

The journey of harnessing electricity began centuries ago. Early pioneers such as Benjamin Franklin, who studied static electricity, and Alessandro Volta, who invented the first battery, laid the foundation for our understanding of electric charge and flow. Later, Michael Faraday's discovery of electromagnetic induction made it possible to generate electricity on a large scale, leading to the development of electric generators.

Electricity is more than just a phenomenon; it is the backbone of modern life. It powers everything from the lights in your home to the smartphones in your hands. It is the driving force behind technological innovations, such as electric vehicles, renewable energy systems, and medical devices that save lives. Without electricity, the conveniences and advancements we enjoy today would not exist.

To fully appreciate electricity, it's important to understand the three key elements that govern its behavior: voltage, current, and resistance. These concepts, explored in the next sections, will provide the foundation for working with electrical systems. By mastering these basics, you'll begin to see how electricity can be controlled, manipulated, and applied to create functional systems that shape the world around us.

Electricity is everywhere, from the natural static charge in the air during a thunderstorm to the carefully designed circuits in your computer. By understanding what electricity is and how it works, you open the door to countless possibilities, from building your first circuit to contributing to groundbreaking advancements in technology.

1.2 Voltage, Current, and Resistance: The Big Three

Voltage, current, and resistance are the foundational concepts in understanding electricity. Together, these three elements define how electrical systems work and how energy flows through circuits. Often referred to as the "big three" of electricity, they are interconnected by a fundamental relationship known as Ohm's Law. Let's explore each concept in detail.

Voltage: The Driving Force
Voltage, also called electric potential difference, is the force that pushes electrons through a circuit. It can be thought of as the pressure in a pipe that drives water to flow. Without voltage, electrons wouldn't move, and no electrical current would exist. Voltage is measured in volts (V), named after Alessandro Volta, who invented the first chemical battery.

Voltage is created by a difference in charge between two points. For example, a battery has a positive terminal (with a deficiency of electrons) and a negative terminal (with an excess of electrons). This difference creates an electric potential, urging electrons to flow from the negative terminal to the positive one when a circuit is connected.

In real-world applications, voltage levels vary significantly. A small flashlight might use a 1.5-volt battery, while household outlets typically provide 120 or 240 volts, depending on the country. High-voltage lines, used to transport electricity over long distances, can carry thousands of volts to minimize energy loss during transmission.

Current: The Flow of Electrons
Electric current is the flow of electrons through a conductor, such as a copper wire. Think of current as the amount of water flowing through a pipe. The more electrons moving through the circuit, the higher the current. Current is measured in amperes (A), often shortened to "amps," and is named after André-Marie Ampère, a pioneer in electromagnetism.

There are two types of current: direct current (DC) and alternating current (AC). In DC, electrons flow in a single direction, as in batteries. In AC, the direction of electron flow alternates periodically, typically 50 or 60 times per second (measured in hertz). AC is the standard for household power because it's more efficient for transmitting electricity over long distances.

Current can be controlled in circuits to power devices of varying sizes and functions. For instance, a small LED might require only a few milliamps (mA), while large motors or appliances may need several amps to operate.

Resistance: The Opposition to Flow
Resistance is the property of a material that opposes the flow of electric current. Imagine a narrow or rough pipe slowing down the flow of water—resistance works in a similar way for electrons. Resistance determines how easily current can flow through a material and is measured in ohms (Ω), named after Georg Simon Ohm, who formulated Ohm's Law.

Different materials have varying levels of resistance. Conductors like copper and silver have low resistance, allowing electricity to flow easily, while insulators like rubber or plastic have high resistance, blocking the flow of current. Resistance in a circuit can also be deliberately added using components like resistors, which help control the amount of current reaching different parts of the circuit.

Factors such as material, temperature, and size influence resistance. For example, longer wires have more resistance than shorter ones, and higher temperatures increase resistance in most materials.

The Interconnection: Ohm's Law
Voltage, current, and resistance are interconnected by Ohm's Law, which states:

$$V = I \times R$$

Where:

- V is voltage (in volts)
- I is current (in amps)

- R is resistance (in ohms)

This equation forms the backbone of electrical circuit calculations. If you know any two of the three variables, you can calculate the third. For instance, if a circuit has a resistance of 10 ohms and a voltage of 20 volts, the current can be calculated as
$I = V/R = 20/10 = 2 \text{ amps}.$

Ohm's Law not only simplifies calculations but also helps in designing circuits to ensure they operate safely and efficiently. For example, knowing the current flowing through a circuit allows you to choose appropriate wire sizes and protective components, like fuses or circuit breakers.

Real-World Examples
To visualize these concepts, consider a simple circuit with a 9-volt battery connected to a small lightbulb. The voltage from the battery drives the current through the circuit, lighting the bulb. The bulb itself provides resistance, which controls the current and ensures the bulb doesn't burn out. If the resistance is too low (e.g., using a larger bulb), the battery might overheat due to excessive current.

Another example is your home's electrical system. Voltage from the power grid is constant, but devices like lamps and refrigerators have varying resistance, which determines the current they draw. High-resistance devices, like LED lights, consume less current, making them energy-efficient.

Voltage, current, and resistance are the foundational building blocks of electricity. Voltage provides the driving force, current represents the flow of electrons, and resistance controls that flow. Together, they define how electricity behaves in circuits, and understanding their interplay is essential for anyone working with electrical systems. Mastering these concepts is your first step toward confidently exploring the world of electrical engineering.

1.3 Ohm's Law and Basic Calculations

Ohm's Law is one of the most fundamental principles in electrical engineering. It provides a mathematical relationship between voltage (V), current (I), and resistance (R) in an electrical circuit. As discussed in the previous section, the formula for Ohm's Law is:

V=I x R

This equation allows us to calculate any one of the three variables if the other two are known. Ohm's Law not only simplifies circuit analysis but also provides insight into how electricity behaves in practical systems. Let's dive deeper into its application and the methods used to solve basic calculations.

Using Ohm's Law in Simple Circuits

Consider a simple circuit consisting of a power source (e.g., a battery), a resistor, and a wire connecting them. If the power source provides a voltage of 9 volts (V) and the resistor has a resistance of 3 ohms (R), you can calculate the current flowing through the circuit using Ohm's Law:

$$I = \frac{V}{R} = \frac{9}{3} = 3 \text{ amps}$$

This tells us that 3 amps of current flow through the circuit. If you know the current and resistance, you can rearrange the formula to find the voltage. Similarly, if you know the voltage and current, you can calculate the resistance:

$$R = \frac{V}{I}$$

These straightforward calculations are the building blocks for analyzing more complex circuits.

Power in Electrical Circuits

In addition to voltage, current, and resistance, power is a key concept in electrical engineering. Power (P) is the rate at which electrical energy is consumed or generated. It is calculated using the formula:

P=V x I

For example, if a device operates at 12 volts and draws 2 amps of current, the power consumption is:

P=12 x 2=24 watts

You can also express power in terms of resistance using variations of Ohm's Law:

$$P = I^2 \times R \quad \text{or} \quad P = \frac{V^2}{R}$$

These formulas are particularly useful for determining how much heat a resistor will produce or evaluating the efficiency of a circuit.

Solving Real-World Problems

Let's explore a practical example to see how Ohm's Law and power calculations work together. Imagine you have a 12-volt car battery connected to a 6-ohm headlight. Using Ohm's Law, you can determine the current flowing through the headlight:

$$I = \frac{V}{R} = \frac{12}{6} = 2 \text{ amps}$$

Next, you can calculate the power consumed by the headlight:

P=V x I=12×2=24 watts

This tells you that the headlight uses 24 watts of power. By understanding these relationships, you can design and troubleshoot circuits effectively.

Parallel and Series Circuits

In more complex circuits, resistors can be arranged in series or parallel configurations. Each arrangement affects the total resistance differently.

- **Series Circuits:** When resistors are connected in series, their resistances add together: $R_{total} = R_1 + R_2 + R_3 + \ldots$

For example, if you have three resistors with values of 2 ohms, 3 ohms, and 5 ohms in series, the total resistance is:
Rtotal = 2+3+5 = 10 ohms

- **Parallel Circuits:** In parallel circuits, the total resistance is calculated using the formula:

$$\frac{1}{R_{total}} = \frac{1}{R_1} + \frac{1}{R_2} + \frac{1}{R_3} + \ldots$$

For instance, if you have two resistors of 6 ohms and 12 ohms in parallel:

$$\frac{1}{R_{total}} = \frac{1}{6} + \frac{1}{12} = \frac{2}{12} + \frac{1}{12} = \frac{3}{12}$$
$$R_{total} = \frac{12}{3} = 4\,\text{ohms}$$

These rules are critical for designing circuits with the desired resistance and current flow.

Applications and Limitations

Ohm's Law is incredibly versatile but has limitations. It applies only to linear circuits where resistance remains constant regardless of voltage or current. Some materials, like semiconductors, have resistance that changes based on conditions like temperature or applied voltage. For such cases, advanced techniques are required.

Understanding and applying Ohm's Law is essential for anyone working with electrical systems. By mastering these calculations, you can predict how circuits will behave, troubleshoot issues, and design systems that operate efficiently and safely. This knowledge serves as the foundation for tackling more complex topics in electrical engineering.

1.4 Electric Circuits: Series vs. Parallel

Electric circuits come in two main configurations: series and parallel. Each type of circuit has unique characteristics that affect how current flows, how voltage is distributed, and how components behave. Understanding these differences is essential for designing and analyzing circuits.

Series Circuits

In a series circuit, all components are connected end-to-end, forming a single path for current to flow. Because the current has only one path, the same amount of current flows through every component in the circuit.

Key Characteristics of Series Circuits:

1. **Current is Constant:** As mentioned earlier, the current (I) remains the same throughout the circuit.
2. **Voltage Splits Across Components:** The total voltage of the power source is divided among the components. For example, if a 12-volt battery is connected to three resistors, the voltage across each resistor depends on its resistance. Using Ohm's Law, the voltage across a single resistor can be calculated.
3. **Total Resistance Adds Up:** The total resistance in a series circuit is the sum of the individual resistances:
   $$R_{total} = R_1 + R_2 + R_3 + \ldots$$

For example, if you have three resistors of 2, 4, and 6 ohms in series, the total resistance is:
Rtotal = 2+4+6 =12 ohms

Advantages and Disadvantages of Series Circuits:
Series circuits are simple to design and build, but they have limitations. If one component fails or is removed, the entire circuit is broken, as the single path for current is interrupted. Additionally, adding more components increases the total resistance, which can reduce the current and limit the circuit's performance.

Example of a Series Circuit:
Imagine a string of old-fashioned holiday lights. If one bulb burns out, the entire string goes dark because the circuit is broken.

Parallel Circuits

In a parallel circuit, components are connected across the same two points, creating multiple paths for current to flow. Unlike series circuits, the current can take different routes, depending on the resistance of each path.

Key Characteristics of Parallel Circuits:

1. **Voltage is Constant Across All Branches:** Each branch of a parallel circuit receives the same voltage as the power source. If a battery provides 9 volts, every branch in the parallel circuit will also experience 9 volts.
2. **Current Splits Across Branches:** The total current (I_{total}) supplied by the power source is divided among the branches. The amount of current in each branch depends on its resistance, calculated using Ohm's Law.
3. **Total Resistance Decreases:** The total resistance in a parallel circuit is always less than the resistance of the smallest branch. It's calculated using the formula:

$$\frac{1}{R_{total}} = \frac{1}{R_1} + \frac{1}{R_2} + \frac{1}{R_3} + \ldots$$

For instance, if you have two resistors of 4 ohms and 6 ohms in parallel:

$$\frac{1}{R_{total}} = \frac{1}{4} + \frac{1}{6} = \frac{3}{12} + \frac{2}{12} = \frac{5}{12}$$

$$R_{total} = \frac{12}{5} = 2.4 \text{ ohms}$$

Advantages and Disadvantages of Parallel Circuits:
Parallel circuits offer redundancy, meaning if one branch fails, the others continue to function. This is why most home wiring systems are parallel circuits—turning off one light doesn't shut off the entire house. However, designing and calculating parallel circuits can be more complex than series circuits, and the increased current demands may require careful planning to avoid overloading the power source.

Example of a Parallel Circuit:
Think of the outlets in your home. Each outlet is connected in parallel, so plugging in or unplugging a device at one outlet doesn't affect the operation of devices at other outlets.

Comparing Series and Parallel Circuits

The choice between series and parallel configurations depends on the application:

- **Series Circuits:** Best for simple applications where maintaining the same current is essential, such as small battery-powered devices.
- **Parallel Circuits:** Ideal for complex systems where independent operation of components is required, such as home electrical systems or vehicle lighting.

Combination Circuits:
Many real-world systems combine series and parallel configurations to balance performance and reliability. For instance, in a car's electrical system, headlights might be connected in

parallel to ensure they operate independently, while dashboard indicators could be wired in series for simplicity.

Practical Applications

In designing circuits, understanding the differences between series and parallel configurations helps optimize performance. For example, LED strip lights often use parallel connections to ensure each section continues to work even if one LED fails. Meanwhile, devices like flashlights use series circuits to ensure all components receive the same current.

As we've discussed in this section and the one before, series and parallel circuits behave differently in terms of current, voltage, and resistance. Series circuits are simple but less robust, while parallel circuits offer flexibility and reliability. Mastering these configurations is crucial for designing circuits that meet specific requirements, laying the groundwork for more advanced topics in electrical engineering.

1.5 Power and Energy in Circuits

Power and energy are two critical aspects of electrical circuits that determine how much work a circuit can perform and how efficiently it operates. Power describes the rate at which energy is used, while energy refers to the total amount of work done over time. These concepts are essential for understanding how circuits function and for designing systems that meet practical needs.

Defining Power in Electrical Circuits

As introduced in previous sections, electrical power (P) is the rate at which electrical energy is consumed or generated in a circuit. It is measured in watts (W), named after James Watt, who made significant contributions to the study of power. The formula for power is: $P = V \times I$, where:

- P is power in watts

- V is voltage in volts
- I is current in amperes

For example, if a device operates at 10 volts and draws 2 amps of current, the power consumed is:
P=10×2=20 watts, this means the device uses 20 watts of energy per second to function.

In cases where only resistance and voltage or current are known, power can also be calculated using variations of Ohm's Law:

$$P = I^2 \times R \quad \text{or} \quad P = \frac{V^2}{R}$$

Energy in Circuits

Electrical energy (E) is the total amount of work done by a circuit over a period of time. It is measured in joules (J) or watt-hours (Wh). The relationship between power and energy is straightforward: E=P x t, where:

- E is energy in joules (or watt-hours for larger scales)
- P is power in watts
- t is time in seconds

For instance, a 60-watt lightbulb running for one hour consumes:
E = 60 W x 3600 s = 216,000 J or 60 Wh

In household applications, energy is typically measured in kilowatt-hours (kWh), where 1 kWh equals 1,000 watt-hours. Utility companies charge consumers based on the number of kilowatt-hours they use.

Applications of Power and Energy

Power and energy play a critical role in selecting components for circuits and designing systems. For example, if you're choosing a resistor for a circuit, you need to ensure that it can handle the power dissipated without overheating. The resistor's power rating must exceed the calculated power to ensure safety and reliability.

In larger systems, such as motors or household appliances, understanding energy consumption helps improve efficiency and reduce costs. For example, energy-efficient devices like LED lights consume significantly less power than incandescent bulbs, saving energy over time.

Power in Series and Parallel Circuits

As we discussed in earlier sections, circuits can be configured in series or parallel, and the arrangement affects how power is distributed.

- **In Series Circuits:** The current is the same through all components, so power consumption depends on the resistance of each component.

For example, in a series circuit with a 10-ohm and a 20-ohm resistor connected to a 30-volt source, the total resistance is $R_{total} = 10 + 20 = 30\,\Omega$, and the current is $I = \frac{V}{R_{total}} = \frac{30}{30} = 1$ A. The power for each resistor is then:

$P_{10\Omega} = I^2 \times R = 1^2 \times 10 = 10\,\text{W}$
$P_{20\Omega} = 1^2 \times 20 = 20\,\text{W}$

- **In Parallel Circuits:** Each branch receives the same voltage, so the power depends on the current through each branch. For example, if a 12-volt battery is connected to two parallel branches with resistances of 6 ohms and 12 ohms, the power for each branch is calculated individually:

$I_{6\Omega} = \frac{V}{R} = \frac{12}{6} = 2\,\text{A} \quad \Rightarrow \quad P = V \times I = 12 \times 2 = 24\,\text{W}$
$I_{12\Omega} = \frac{12}{12} = 1\,\text{A} \quad \Rightarrow \quad P = 12 \times 1 = 12\,\text{W}$

Practical Considerations

When designing circuits, managing power is crucial for safety and efficiency. Overloading a circuit by exceeding its power capacity can cause components to overheat or fail, potentially leading to dangerous situations like fires.

To prevent issues, electrical systems include protective devices like fuses and circuit breakers, which interrupt the circuit when the current exceeds safe limits. Proper wire sizing is also important, as wires that are too thin may overheat under high current loads.

Energy Efficiency

Energy efficiency is a key consideration in modern electrical systems. Devices like solar panels and energy-efficient appliances are designed to minimize energy waste. For instance, an inverter in a solar power system converts DC electricity from solar panels into usable AC power while reducing energy losses. Similarly, energy-efficient motors and lighting systems reduce power consumption without compromising performance.

Conclusion

Power and energy are essential for evaluating how circuits operate and perform. Power indicates the rate at which energy is used, while energy measures the total work done over time. These concepts help determine efficiency, safety, and cost in both simple and complex systems. By applying key formulas and understanding their real-world implications, you can design circuits that are not only functional but also optimized for performance. Mastering power and energy provides a solid foundation for tackling advanced electrical engineering challenges.

1.6 Safety Precautions When Working with Electricity

Working with electricity requires more than just understanding circuits; it demands a respect for the power and risks involved. While electricity is a tool that drives innovation and convenience, it also poses significant dangers if handled carelessly. Following essential safety precautions not only protects you from injury but also ensures the longevity and functionality of your projects.

Why Electrical Safety Matters

Electricity moves invisibly and instantaneously, leaving little room for error. Even low-voltage circuits can cause shocks or burns, while high-voltage systems present life-threatening risks. As simple as turning off the power might seem, overlooking even basic precautions can lead to accidents. Recognizing potential hazards and taking deliberate steps to mitigate them is the foundation of safe electrical work.

Start by Controlling the Power

The first rule of working with electricity is to ensure the circuit is not live. Always disconnect the power source before starting any work. This could mean unplugging the device, switching off the breaker, or removing batteries. If the circuit stores energy, as capacitors do, discharge them safely before proceeding. As noted earlier, verifying the absence of voltage using a multimeter or voltage tester is crucial. Assume a circuit is live until proven otherwise.

Personal Protection: More Than Just Gloves

The right gear can mean the difference between a minor mishap and a serious injury. Insulating gloves and rubber-soled shoes are essential for creating a barrier between you and electrical currents. Safety goggles protect your eyes from potential sparks or debris, especially during soldering or when working with older, brittle wires. Beyond equipment, avoid wearing metal accessories like rings or watches that can conduct electricity.

A common mistake is underestimating the importance of clothing. Loose-fitting garments can catch on wires or tools, while synthetic fabrics may melt and stick to your skin in the event of a fault. Opt for fitted, non-flammable clothing to reduce these risks.

The Workspace: Your First Line of Defense

A clean and organized workspace minimizes accidents. Keep wires, tools, and components neatly arranged to prevent tangles and short circuits. Ensure your workspace is dry, as even a small amount of water can significantly increase the risk of shock. If you're working outdoors or in humid environments, take extra precautions like using ground fault circuit interrupters (GFCIs) to cut off power instantly in case of a fault.

Fire safety is another critical consideration. Always have a fire extinguisher nearby, rated for electrical fires. Remember, never use water to extinguish an electrical fire, as it can worsen the situation.

Common Hazards and How to Avoid Them

One of the most frequent causes of electrical accidents is damaged equipment. Frayed wires, cracked insulation, or loose connections can lead to sparks, overheating, or short circuits. Inspect all components thoroughly before use, and replace any damaged items immediately.

Overloading circuits is another avoidable hazard. Plugging too many devices into a single outlet or using an undersized wire for a high-current application can generate excessive heat. Using fuses or circuit breakers designed for the circuit's capacity helps prevent overloads and the resulting risks of fire or equipment failure.

Approaching High-Risk Components

Batteries and capacitors deserve special attention due to their ability to store energy. A charged capacitor, even in a powered-off circuit, can deliver a sudden, dangerous shock. Discharge capacitors carefully using a resistor, and always follow manufacturer guidelines for handling high-capacity units. Similarly, batteries, especially lithium-ion types, can overheat or even explode if short-circuited. Handle them with care and avoid physical damage to their casing.

Soldering Safely

For many hobbyists and beginners, soldering is a key part of working with circuits. While it's a valuable skill, it comes with its own set of risks. A soldering iron operates at extremely high temperatures, so always use its stand when not actively soldering. Work in a well-ventilated area to avoid inhaling fumes, which can be harmful over time. After soldering, inspect the joints to ensure there are no unintentional connections that could cause a short circuit.

Emergency Preparedness

No matter how careful you are, accidents can happen. Knowing how to respond is just as important as prevention. If someone receives an electric shock, never touch them directly; instead, disconnect the power and use a non-conductive object to separate them from the circuit. Call emergency services immediately and provide first aid only if you are trained to do so.

Similarly, in the case of an electrical fire, evacuate the area if the fire grows beyond your control. Use a proper fire extinguisher to address smaller incidents safely. Having an emergency plan in place for these situations ensures you can act quickly and effectively.

The Role of Awareness and Vigilance

Safety isn't just about following a set of rules—it's a mindset. Always remain aware of your surroundings and think critically about the potential risks of each action. For beginners, this means starting with low-voltage projects and gradually advancing to more complex circuits as your confidence and knowledge grow. By practicing vigilance and incorporating safety into every step of your work, you not only protect yourself but also build good habits that will carry over into future projects. Electrical engineering is as much about caution as it is about creativity, and prioritizing safety ensures that your learning journey remains productive and incident-free.

Chapter 2: Understanding Electrical Components

2.1 Resistors, Capacitors, and Inductors

Resistors, capacitors, and inductors are the fundamental building blocks of electrical circuits. Each plays a distinct role, shaping how current and voltage behave within a system. Together, these components enable engineers to control, store, and modify electrical energy to achieve desired outcomes in circuits.

Resistors: Controlling Current and Voltage
As the name suggests, resistors resist the flow of electric current. They are used to control the amount of current in a circuit and divide voltage between components. Their resistance, measured in ohms (Ω), determines how much they oppose current flow.

Resistors come in various forms, including fixed resistors, which have a set resistance value, and variable resistors, such as potentiometers, which allow the resistance to be adjusted. For example, a potentiometer might be used as a volume control knob in audio equipment.

Applications of Resistors:

- **Voltage Division:** In circuits where a specific voltage is required, resistors create voltage dividers. For example, two resistors in series can split a voltage source into smaller parts.
- **Current Limiting:** Resistors protect components like LEDs by limiting the amount of current flowing through them.
- **Signal Conditioning:** In analog circuits, resistors are often combined with capacitors or inductors to filter or shape signals.

When selecting resistors, power ratings (measured in watts) are as important as resistance. A resistor must handle the heat generated by the current passing through it, calculated using $P = I^2 \times R$.

Capacitors: Storing and Releasing Energy

Capacitors store electrical energy temporarily in an electric field. They consist of two conductive plates separated by an insulating material called a dielectric. When a voltage is applied, charges accumulate on the plates, creating an electric field.

Capacitance, measured in farads (F), indicates a capacitor's ability to store charge. Most everyday capacitors, like those in electronics, have capacitance values in microfarads (μF) or picofarads (pF).

Types of Capacitors:

- **Ceramic Capacitors:** Compact and inexpensive, ideal for high-frequency applications.
- **Electrolytic Capacitors:** Provide high capacitance, commonly used for power supply smoothing.
- **Tantalum Capacitors:** Compact and stable, suitable for precision applications.

Applications of Capacitors:

- **Energy Storage:** Capacitors supply bursts of energy in systems like camera flashes or defibrillators.
- **Filtering:** Capacitors smooth voltage fluctuations in power supplies by filtering out high-frequency noise.
- **Coupling and Decoupling:** They block DC components in signals or stabilize voltage in sensitive circuits.

Capacitors are particularly useful in applications where fast charging and discharging are required, such as signal processing or energy backup systems.

Inductors: Managing Magnetic Fields

Inductors are coils of wire designed to store energy in a magnetic field when current flows through them. Unlike capacitors, which store energy in an electric field, inductors rely on magnetism. The inductance of an inductor, measured in henries (H), depends on the number of coil turns, the core material, and the coil's shape.

Applications of Inductors:

- **Filtering:** Inductors are commonly used in filters to block high-frequency signals while allowing low-frequency signals to pass.
- **Energy Storage:** In switching power supplies, inductors temporarily store energy during voltage conversion.
- **Transformers:** Inductors play a critical role in transformers, which transfer energy between circuits and step voltage up or down.

One key property of inductors is their opposition to changes in current. This behavior, known as inductive reactance, makes them valuable in circuits that need to regulate or stabilize current flow.

Practical Considerations
While resistors, capacitors, and inductors are versatile, they are not without limitations. Resistors dissipate energy as heat, requiring careful selection to prevent overheating. Capacitors degrade over time, especially electrolytic types, and must be replaced periodically. Inductors can be bulky and heavy, particularly in high-power applications, due to the core material.

Working Together
In many circuits, these components are used together to achieve specific functions. For instance, in an RC (resistor-capacitor) circuit, resistors and capacitors work to filter signals or create time delays. Similarly, LC (inductor-capacitor) circuits are essential in radio frequency applications, such as tuning antennas.
Understanding the roles and limitations of resistors, capacitors, and inductors is crucial for designing efficient and reliable circuits.

These fundamental components form the backbone of nearly every electrical system, providing the tools to control, store, and shape electrical energy.

2.2 Diodes and Transistors

Diodes and transistors are essential components in modern electronics, enabling everything from simple signal rectification to complex computational functions. While resistors, capacitors, and inductors control or store energy, diodes and transistors manipulate and amplify the flow of current, making them indispensable in digital and analog circuits.

Diodes: One-Way Valves for Current

A diode is a semiconductor device that allows current to flow in one direction while blocking it in the opposite direction. This directional behavior makes diodes crucial for tasks like rectification, signal processing, and protection circuits.

Structure and Operation
A typical diode consists of a junction between two semiconductor materials: p-type (positive) and n-type (negative). This junction, called a PN junction, creates an internal electric field that only permits current to flow when the diode is forward-biased—meaning the positive voltage is applied to the p-side and the negative voltage to the n-side.

When reverse-biased, the diode blocks current flow, except for a small leakage current. If the reverse voltage exceeds a certain limit, called the breakdown voltage, the diode may conduct uncontrollably, potentially causing damage unless the diode is designed to handle it (e.g., Zener diodes).

Types of Diodes

1. **Standard Rectifier Diodes:** Used to convert AC to DC in power supplies.

2. **Zener Diodes:** Designed to conduct in reverse beyond a specific voltage, commonly used in voltage regulation.
3. **Light-Emitting Diodes (LEDs):** Emit light when current flows through them, widely used in indicators and displays.
4. **Schottky Diodes:** Known for their fast switching and low forward voltage drop, making them ideal for high-frequency circuits.

Applications of Diodes

- **Rectification:** Diodes convert AC to DC in power supplies, as seen in bridge rectifier circuits.
- **Voltage Regulation:** Zener diodes maintain a stable voltage across sensitive components.
- **Signal Protection:** Diodes protect circuits by clamping high voltages or preventing reverse polarity damage.

Transistors: Amplifiers and Switches

Transistors are more complex than diodes and serve as the backbone of modern electronics. They can amplify electrical signals or act as electronic switches, enabling countless applications in computation, communication, and control systems.

Structure and Types

A transistor is a three-layer semiconductor device with three terminals: the base, collector, and emitter (in bipolar junction transistors, or BJTs) or the gate, drain, and source (in field-effect transistors, or FETs). The two most common types of transistors are:

1. **Bipolar Junction Transistors (BJTs):** These have three regions—emitter, base, and collector—and are classified as NPN or PNP based on their doping arrangement.

2. **Field-Effect Transistors (FETs):** These control current using an electric field and are categorized into junction FETs (JFETs) and metal-oxide-semiconductor FETs (MOSFETs). MOSFETs are widely used in modern digital circuits due to their high efficiency and low power consumption.

How Transistors Work

In a BJT, a small current at the base controls a much larger current between the collector and emitter. For instance, in an NPN transistor, applying a positive voltage to the base allows current to flow from the collector to the emitter.

In contrast, FETs use voltage to control the flow of current. For example, in a MOSFET, applying a voltage to the gate terminal creates an electric field that modulates the current flow between the source and drain.

Applications of Transistors

1. **Amplification:** Transistors amplify weak signals in audio devices, radios, and communication systems. For example, in a microphone circuit, a transistor boosts the tiny electrical signal generated by sound waves.
2. **Switching:** Transistors serve as electronic switches in digital logic circuits, turning current flow on or off based on input signals. This functionality underpins processors, memory chips, and microcontrollers.
3. **Voltage Regulation:** Transistors are key components in voltage regulators and power management circuits.
4. **Oscillators:** Transistors create oscillations in circuits, which are essential for generating radio frequencies in communication devices.

Diodes and Transistors in Combination

Diodes and transistors often work together in circuits to achieve specific goals. For instance, in a rectifier circuit, diodes convert AC to DC, while a transistor stabilizes the output voltage. In digital logic circuits, transistors form the core of logic gates, while diodes protect against voltage spikes.

Selecting the Right Component

Choosing the appropriate diode or transistor for a circuit depends on several factors, including voltage and current requirements, frequency, and power dissipation. For example, Schottky diodes are ideal for high-speed switching, while MOSFETs are preferred for high-efficiency power applications.

Building Blocks of Modern Electronics

While resistors, capacitors, and inductors manage basic electrical behaviors, diodes and transistors bring functionality and intelligence to circuits. They allow engineers to shape signals, amplify information, and enable automation. Understanding their properties and how to apply them is essential for tackling more advanced designs and projects.

2.3 Integrated Circuits

Integrated circuits (ICs) are among the most transformative innovations in electronics, enabling the development of compact, powerful, and reliable devices. An integrated circuit is a single semiconductor chip that combines multiple components, such as transistors, diodes, resistors, and capacitors, into a unified package. This miniaturization allows circuits that once occupied entire rooms to fit onto a fingernail-sized chip.

What Are Integrated Circuits?

Integrated circuits are built on a wafer of silicon, a semiconductor material. Using photolithography and other advanced manufacturing techniques, various electronic components are etched and deposited onto the silicon to create an interconnected network. Once the wafer is processed, it is cut into individual chips and encased in protective packaging with external pins or leads for connection to other components.

There are two primary types of ICs:

1. **Analog ICs:** These process continuous signals, such as amplifying weak audio signals or managing power in voltage regulators.
2. **Digital ICs:** These handle binary data (0s and 1s) and are the foundation of processors, memory, and logic gates.

Some ICs, known as **mixed-signal ICs**, combine analog and digital functionalities, making them ideal for complex applications like smartphones and communication systems.

Advantages of Integrated Circuits

Integrated circuits revolutionized electronics for several reasons:

- **Miniaturization:** ICs drastically reduce the size of electronic systems, allowing portable and compact devices.
- **Cost Efficiency:** Mass production techniques make ICs cheaper than assembling individual components.
- **Reliability:** Fewer physical connections and compact design reduce the risk of mechanical failures.
- **Power Efficiency:** ICs consume less power compared to discrete components performing the same functions.

Common Types of ICs

Integrated circuits come in various forms, each tailored to specific applications:

1. Microprocessors:
Microprocessors are a type of IC that serve as the brain of computers and other devices. They execute instructions, perform

calculations, and manage data flow. For example, the central processing unit (CPU) in your computer or the microcontroller in a washing machine are powered by microprocessors.

2. Operational Amplifiers (Op-Amps):
Op-amps are analog ICs used for signal processing tasks like amplification, filtering, and comparison. For instance, an op-amp in a stereo amplifier boosts weak audio signals to drive speakers.

3. Memory ICs:
Memory ICs store data in digital systems. They include volatile memory, like RAM, for temporary storage, and non-volatile memory, like flash, for permanent storage in devices such as USB drives and SSDs.

4. Application-Specific Integrated Circuits (ASICs):
ASICs are custom-designed ICs tailored to perform specific tasks. For example, an ASIC in a smartphone might optimize its performance for image processing or wireless communication.

5. Power Management ICs (PMICs):
These ICs regulate and distribute power within a system. They are essential in energy-efficient designs, such as battery-powered devices.

How Integrated Circuits Are Used

ICs are the backbone of modern electronics. Here are some examples of their applications:

- **Consumer Electronics:** ICs are found in nearly every device, including smartphones, televisions, and gaming consoles.
- **Industrial Systems:** ICs control machinery, manage sensors, and process data in manufacturing and automation.

- **Medical Devices:** Pacemakers, diagnostic equipment, and wearable health monitors depend on ICs for their functionality.
- **Communication Systems:** ICs enable high-speed data transmission in mobile networks, Wi-Fi routers, and satellite communication.

Limitations of Integrated Circuits

Despite their advantages, ICs have limitations:

- **Thermal Sensitivity:** Excessive heat can damage ICs, requiring careful thermal management, such as heat sinks or cooling fans.
- **Repair Challenges:** Unlike discrete components, ICs cannot be repaired if a single component inside fails. The entire chip must be replaced.
- **Design Complexity:** Creating ICs requires advanced tools and expertise, making development costly and time-consuming.

The Future of Integrated Circuits

Advancements in IC technology continue to push the boundaries of what is possible. Modern ICs incorporate billions of transistors, as seen in processors for high-performance computing. Emerging technologies, like quantum computing and neuromorphic chips, aim to revolutionize how we process and store information. Additionally, flexible and biodegradable ICs are being developed for applications in wearable tech and sustainable electronics.

Integrated circuits have been a driving force behind the digital age, enabling everything from personal devices to global communication networks. As IC technology evolves, it promises to further transform industries and pave the way for innovations we've yet to imagine. Understanding ICs is essential for grasping the full scope of modern electronics and their future potential.

2.4 Switches, Relays, and Connectors

Switches, relays, and connectors might not seem as exciting as transistors or integrated circuits, but they are indispensable for making electrical systems work smoothly. These components enable control, automation, and connectivity in circuits, allowing you to interact with devices, manage power, and connect components easily and safely.

Think of a switch as the simplest way to control electricity. Whether it's flipping a light switch or pressing a button to turn on a device, switches are everywhere in our daily lives. Relays take this concept a step further, allowing electrical signals to control circuits automatically. And connectors? They're the unsung heroes that ensure everything stays physically and electrically connected, simplifying assembly and repair.

Switches: Simple but Essential

At its core, a switch is just a device that opens or closes a circuit, controlling the flow of current. But even something so straightforward comes in a variety of designs, each suited for specific tasks.

Consider the light switch on your wall. It's a **single-pole single-throw (SPST)** switch, the simplest kind. Flip it one way, and the circuit is complete; flip it the other way, and the circuit is broken. On the other hand, if you've ever used a three-way switch to control a light from two locations, you've encountered a **single-pole double-throw (SPDT)** switch. These switches can toggle between two outputs, giving you more control over how current flows.

Switches also come in other forms, like **push buttons** on keyboards or remote controls, and **rotary switches** in devices like ovens or old-school stereos. Each type is tailored to its role, but they all perform the same basic function: letting you decide when and how electricity flows.

Relays: Automation in Action

While switches require you to physically interact with them, relays do the same job electrically. Imagine a situation where you need to turn on a heavy-duty motor but don't want to handle high currents directly. That's where a relay shines. A small control signal—often just a low-current circuit—activates the relay, which then switches on the high-current circuit safely.

Relays use an electromagnet to open or close their contacts. When current flows through the coil, it generates a magnetic field that pulls the contacts into position. This design makes relays incredibly versatile. For example, they're used in car starters, industrial machinery, and even home automation systems.

Modern versions, like **solid-state relays (SSRs)**, have no moving parts, relying on electronic components to perform the switching. This makes them faster and longer-lasting, especially in high-frequency applications. But for everyday tasks, traditional electromechanical relays remain reliable workhorses.

Connectors: Keeping It All Together

If switches and relays control how circuits operate, connectors ensure that everything stays linked together. Whether you're plugging in a power cord or snapping a data cable into a port, connectors are the physical bridges between components.

Think of the USB port on your laptop. It's a **plug-and-socket connector**, designed for easy insertion and removal. Inside your computer, though, you'll find more specialized connectors, like **header pins** that attach components to the motherboard. In industrial settings, **terminal blocks** provide a secure and durable way to connect wires, while **coaxial connectors** ensure high-speed data transmission in cable systems.

Choosing the right connector depends on the job. For high-power circuits, you need connectors rated for the current they'll carry. In outdoor or harsh environments, weatherproof connectors protect against water and dust. And in applications where components might need frequent replacement, modular connectors make disassembly a breeze.

How They Work Together

Switches, relays, and connectors often team up to create functional and flexible systems. Picture this: a wall-mounted switch controls a relay in your home's HVAC system, turning on a high-power motor remotely. The relay, in turn, connects to the motor via robust industrial connectors that can handle the load. Meanwhile, the control panel might have modular connectors to allow easy maintenance or upgrades.

This collaboration makes systems reliable and user-friendly. Without connectors, circuits would be a tangled mess of soldered wires. Without relays, we'd need to handle dangerous high-current loads directly. And without switches, we'd have no way to interact with our devices.

Everyday Dependability

It's easy to overlook these components because they're so ubiquitous. But the next time you press a button, plug in a cable, or hear the faint click of a relay, you're witnessing the quiet but critical role switches, relays, and connectors play in our lives. They're not just hardware—they're enablers of convenience, safety, and efficiency in every electrical system we use.

2.5 Batteries and Power Supplies

Batteries and power supplies are the lifeblood of electrical systems. Without them, circuits would be lifeless, unable to function as designed. Batteries provide portable, stored energy, while power supplies deliver stable and regulated energy to devices. Understanding how these components work and their differences is crucial for building or troubleshooting electrical systems.

Batteries: Stored Energy on Demand

Batteries are electrochemical devices that convert chemical energy into electrical energy. They are widely used because of their portability, making them ideal for everything from flashlights to electric vehicles.

How Batteries Work
A battery consists of one or more cells. Each cell contains three main parts:

1. **Anode (Negative Terminal):** Where oxidation occurs, releasing electrons.
2. **Cathode (Positive Terminal):** Where reduction takes place, accepting electrons.
3. **Electrolyte:** A medium that allows ions to move between the anode and cathode, completing the chemical reaction.

When the battery is connected to a circuit, electrons flow from the anode to the cathode through the external circuit, powering the connected device.

Types of Batteries
Batteries come in two main categories:

- **Primary Batteries:** Non-rechargeable and designed for single use. Common examples include alkaline batteries (used in remote controls) and lithium primary batteries (used in smoke detectors).
- **Secondary Batteries:** Rechargeable and designed for repeated use. Examples include lithium-ion batteries (used in smartphones), lead-acid batteries (common in cars), and nickel-metal hydride (NiMH) batteries (found in rechargeable household items).

Key Parameters of Batteries

- **Capacity:** Measured in milliampere-hours (mAh) or ampere-hours (Ah), this indicates how much energy a battery can store.
- **Voltage:** Determined by the battery's chemistry, with common values like 1.5V (alkaline) or 3.7V (lithium-ion).
- **Discharge Rate:** The speed at which a battery can deliver current without overheating or degrading.

Applications of Batteries

Batteries are versatile and used across industries:

- **Portable Electronics:** Smartphones, laptops, and cameras rely on lightweight lithium-ion batteries.
- **Electric Vehicles (EVs):** Large battery packs power EVs, storing enough energy for hundreds of miles.
- **Backup Systems:** Lead-acid batteries are often used in uninterruptible power supplies (UPS) to maintain power during outages.

Power Supplies: Stable Energy for Devices

While batteries are portable, power supplies are stationary systems that provide continuous energy to devices connected to an external source, such as the electrical grid. Power supplies ensure circuits receive a consistent and appropriate voltage, regardless of fluctuations in the source.

Types of Power Supplies

1. **Linear Power Supplies:** These provide a stable DC voltage using a transformer and rectifier. They are simple but less efficient because they dissipate excess energy as heat.
2. **Switching Power Supplies (SMPS):** More efficient than linear supplies, SMPS use high-frequency switching and voltage regulation to convert power. They are compact and widely used in modern electronics, like laptops and chargers.

3. **Uninterruptible Power Supplies (UPS):** These combine a battery and an inverter to provide backup power during grid outages. UPS systems are critical for data centers and medical equipment.
4. **Programmable Power Supplies:** Advanced supplies allow users to set specific voltage and current levels, often used in testing and development environments.

Key Parameters of Power Supplies

- **Voltage Regulation:** Ensures the output voltage remains stable even if the input voltage fluctuates.
- **Current Capacity:** Indicates the maximum current the supply can deliver without overheating.
- **Efficiency:** Measures how effectively the power supply converts input energy to usable output energy.

Applications of Power Supplies

Power supplies are critical in virtually all electronic systems:

- **Consumer Electronics:** They power household devices like televisions, game consoles, and routers.
- **Industrial Equipment:** Machines often require custom power supplies to meet high current demands.
- **Laboratory Use:** Bench power supplies provide precise and adjustable power for testing circuits.

Choosing Between Batteries and Power Supplies

Deciding whether to use a battery or a power supply depends on the application's requirements:

- **Portability Needs:** Batteries are ideal for portable devices where grid access isn't available.
- **Energy Stability:** Power supplies are better for systems requiring uninterrupted and stable energy.

- **Rechargeability:** While some batteries are rechargeable, they may not meet the continuous demands of high-power systems, making power supplies a better choice.

Battery and Power Supply Safety

Both batteries and power supplies come with safety concerns:

- **Batteries:** Improper handling of batteries, especially lithium-ion types, can lead to short circuits, overheating, or even explosions. Always store batteries in a cool, dry place and avoid physical damage to their casing.
- **Power Supplies:** Overloading or poor ventilation can cause overheating and failure. Ensure power supplies are rated for the devices they power, and never block cooling vents.

The Future of Batteries and Power Supplies

Advancements in energy storage and power delivery are transforming these components. Solid-state batteries, with higher energy densities and improved safety, are set to revolutionize portable devices and electric vehicles. Similarly, renewable energy systems like solar panels are driving innovation in power supplies, leading to more efficient inverters and grid-independent solutions.

Batteries and power supplies form the foundation of modern electrical systems. From powering small gadgets to supporting critical infrastructure, they ensure that energy is always available when and where it's needed. Whether you're building a simple circuit or designing a complex system, understanding these components is essential for success.

2.6 Real-World Examples of Component Usage

Electrical components like resistors, capacitors, diodes, transistors, and others are not just theoretical building blocks—they are everywhere in the devices and systems we interact with daily.

Understanding how these components come together to solve real-world problems can demystify electrical engineering and highlight the practical relevance of what you've learned so far.

Let's explore how these components are applied in real-world devices and systems, focusing on their roles and how they enable functionality.

1. Resistors in Action: Setting Limits and Dividing Voltage

Example 1: LED Circuits
If you've ever built a simple LED circuit, you've encountered a resistor. LEDs are sensitive to current, and exceeding their maximum current rating can cause them to burn out. A resistor placed in series with the LED limits the current to a safe value. For instance, if you power an LED with a 9V battery, you might use a resistor to drop the voltage across the LED to its safe operating range.

Example 2: Volume Controls
Resistors also play a central role in audio systems. A potentiometer, which is a type of variable resistor, is commonly used as a volume control knob. Turning the knob adjusts the resistance in the circuit, which changes the signal strength sent to the speaker, making the audio louder or quieter.

2. Capacitors: Storing and Filtering Energy

Example 1: Power Supply Smoothing
In power supplies, capacitors smooth out fluctuations in voltage. As explained in earlier sections, raw voltage from a rectifier can contain ripples. A capacitor placed across the output of the rectifier charges during voltage peaks and discharges during dips, creating a more stable DC output.

Example 2: Camera Flashes
The quick burst of intense light from a camera flash relies on a capacitor. The capacitor stores energy from a battery and releases

it all at once to the flash bulb. This rapid discharge produces the high-energy pulse needed for the flash.

3. Diodes: One-Way Control

Example 1: Protecting Circuits
Diodes are often used to protect circuits from damage caused by reverse polarity. For example, if you accidentally connect a battery backward, the diode blocks the reverse current, preventing harm to sensitive components.

Example 2: Solar Panels
In solar panel systems, diodes prevent backflow of current when the panels aren't producing power, such as during nighttime. These blocking diodes ensure that energy stored in batteries doesn't drain back into the solar panels.

4. Transistors: Switching and Amplification

Example 1: Microcontroller Outputs
Microcontrollers often lack the ability to drive high-current loads directly. For example, if a microcontroller is controlling a motor, a transistor can act as a switch. The small control signal from the microcontroller activates the transistor, allowing it to handle the high current needed by the motor.

Example 2: Amplifiers
In audio amplifiers, transistors are the key components that boost weak signals from sources like microphones or musical instruments. The amplified signal is then strong enough to drive speakers. Without transistors, modern sound systems would not exist.

5. Inductors: Managing Magnetic Energy

Example 1: Transformers
Inductors are at the heart of transformers, which are used in power distribution. A transformer steps up the voltage for efficient long-

distance transmission and steps it down again for safe use in homes and businesses.

Example 2: Radio Tuners
In radios, inductors combine with capacitors to form LC circuits that can tune into specific frequencies. By adjusting the inductor or capacitor, the circuit resonates at the desired frequency, isolating a particular radio station.

6. Integrated Circuits: Simplifying Complex Systems

Example 1: Microprocessors in Computers
The central processing unit (CPU) in your computer is an integrated circuit that contains billions of transistors. It performs countless calculations per second, enabling everything from word processing to video streaming.

Example 2: Op-Amps in Sensors
Operational amplifiers (op-amps) are ICs used to process signals from sensors. For example, in a temperature monitoring system, an op-amp amplifies the tiny voltage change generated by the temperature sensor, making it readable by a microcontroller.

7. Switches and Relays: Manual and Automatic Control

Example 1: Home Lighting
The light switches in your home are simple SPST switches. They either connect or disconnect the circuit powering your lights. In some cases, smart switches include relays and microcontrollers for remote operation.

Example 2: Automotive Systems
Relays in cars control everything from the headlights to the fuel pump. For example, when you turn on the ignition, a relay ensures that the starter motor gets the high current it needs while protecting the car's control systems from excessive current.

8. Batteries and Power Supplies: Portable and Continuous Power

Example 1: Smartphones
Lithium-ion batteries power most smartphones, providing lightweight, high-capacity energy storage. Charging circuits inside the phone include ICs that regulate voltage and current to safely charge the battery without overheating.

Example 2: Industrial Power Supplies
In factories, machines often require specialized power supplies to deliver stable energy. For example, programmable power supplies allow engineers to test circuits under various voltage and current conditions, ensuring reliability before deploying a system.

9. Connectors: Building Reliable Connections

Example 1: USB Ports
USB connectors are everywhere, enabling quick and easy connections for charging devices or transferring data. Inside your computer, these connectors also ensure communication between the motherboard and peripherals.

Example 2: Aerospace Applications
In aerospace systems, connectors must withstand extreme conditions, including vibration, temperature changes, and high altitudes. Specialized connectors are designed to maintain reliable connections in these harsh environments.

The Bigger Picture: How Components Work Together

No component works in isolation; the magic of electronics happens when components are combined. For example, consider a simple smartphone charger. The power supply converts AC to DC and regulates the output voltage. Inside the phone, resistors, capacitors, and diodes protect and stabilize the charging circuit. Meanwhile, the battery stores energy, and the microcontroller manages charging to prevent overheating.

In more complex systems, like electric vehicles, components work together on a much larger scale. The battery stores energy, transistors control the motors, capacitors filter noise, and microprocessors ensure that the entire system operates efficiently.

Everyday Engineering

From the smallest LED to the most advanced computer, every electronic device you encounter is a testament to the careful selection and combination of components. By understanding these real-world examples, you're not just learning about circuits—you're seeing the immense potential of electrical engineering in action. Each resistor, capacitor, and transistor plays a part in creating systems that power our modern world.

Chapter 3: Fundamentals of Circuit Design

3.1 What Makes a Circuit Work?

At its core, a circuit is a closed loop through which electrical current flows. This flow of current is what powers devices, lights LEDs, and drives motors. But what makes a circuit work? The answer lies in how components interact within the loop to guide, control, and utilize electrical energy effectively.

A functional circuit requires three essential elements: a power source, conductive pathways, and components that perform specific tasks. When these elements come together in the right configuration, the circuit performs a desired function, whether it's amplifying a signal, charging a battery, or running a computer processor.

The Role of the Power Source

Every circuit starts with a power source. As discussed in **Chapter 2**, this could be a battery, a power supply, or another energy source. The power source provides the voltage necessary to push electrons through the circuit.

Voltage, often referred to as "electric pressure," creates the potential difference that drives current. Without it, electrons remain stationary, and the circuit doesn't function. For example, in a flashlight, the battery supplies the energy to illuminate the bulb by pushing current through it.

Conductive Pathways: The Circuit's Highways

Conductive materials, usually copper wires or PCB (printed circuit board) traces, create pathways for electrons to flow. These pathways connect the components in a circuit, forming the closed loop needed for current to travel.

Conductors must be chosen carefully to ensure they can handle the current without overheating or introducing too much resistance. Wires that are too thin, for instance, might overheat under high current loads, potentially causing damage or even fire.

Circuit design also requires careful attention to how pathways are arranged. If pathways cross unintentionally or touch in the wrong places, a short circuit can occur, leading to component failure or even damage to the power source.

Components: The Heart of the Circuit

While the power source and conductive pathways set the stage, the components in the circuit perform the actual work. These include:

- **Resistors:** Control the flow of current and adjust voltage levels.
- **Capacitors:** Store and release energy, filter signals, or smooth voltage.
- **Diodes:** Ensure current flows in only one direction.
- **Transistors:** Act as switches or amplifiers.
- **Integrated Circuits (ICs):** Handle complex tasks like signal processing or computation.

Each component has a specific role, and their arrangement within the circuit determines its functionality. For instance, placing a resistor in series with an LED ensures that the current is limited to a safe level, preventing the LED from burning out.

Circuit Configurations: Series and Parallel

As explained in **Chapter 1**, circuits can be configured in series, parallel, or a combination of both. These configurations influence how voltage and current are distributed among the components.

- **Series Circuits:** Current is the same through all components, but voltage is divided. This configuration is

common in applications like old-fashioned Christmas lights.
- **Parallel Circuits:** Voltage is the same across all branches, but current divides based on the resistance of each branch. This is how home wiring systems are typically designed, allowing independent operation of devices.

The choice of configuration depends on the circuit's purpose. For instance, parallel configurations are often used when consistent voltage across multiple components is required.

The Importance of a Closed Loop

For any circuit to function, it must form a complete loop. If the circuit is open—due to a broken connection, an incomplete pathway, or a faulty component—current cannot flow. This principle is fundamental to circuit design and troubleshooting.

Switches are a great example of how this concept is applied. When a switch is "off," it breaks the loop, stopping current flow. When the switch is turned "on," it closes the loop, allowing current to flow through the circuit.

Energy Conversion in Circuits

Circuits are designed to convert electrical energy into other forms of energy. For example:

- **Light Circuits:** LEDs and incandescent bulbs convert electrical energy into light.
- **Sound Circuits:** Speakers convert electrical signals into sound waves.
- **Motion Circuits:** Motors transform electrical energy into mechanical movement.

Each of these conversions relies on the interaction of components and the efficient flow of current through the circuit.

Designing for Functionality

To make a circuit work, it's essential to start with a clear understanding of its purpose. Ask yourself:

- What is the circuit supposed to do?
- What components are needed to achieve that function?
- How should the components be arranged to ensure proper operation?

A simple flashlight circuit, for instance, requires a battery (power source), a bulb (load), and a switch to control the flow of current. More complex designs, like a radio, require additional components like capacitors, transistors, and inductors to process signals and amplify sound.

A Symphony of Interactions

What makes a circuit work is the seamless interaction between its power source, pathways, and components. Each element plays a critical role, and even a single missing connection or faulty component can render the circuit useless. By understanding these fundamentals and the interplay between elements, you can design circuits that not only function but also perform reliably and efficiently. This foundational knowledge sets the stage for exploring tools, techniques, and hands-on design in the sections that follow.

3.2 Basic Tools for Circuit Design and Testing

Designing and testing circuits require more than just theoretical knowledge—you need the right tools to bring your ideas to life and ensure they work as intended. These tools help you assemble components, measure critical parameters, troubleshoot issues, and refine your designs. For beginners and seasoned professionals alike, having a reliable set of tools is essential for success in circuit design.

Prototyping Tools: Bringing Circuits to Life

Before building a final circuit, you'll often start with a prototype to test your ideas. Prototyping tools let you create temporary, modifiable setups to experiment with different configurations.

Breadboards are the most common prototyping tools for beginners. They allow you to connect components without soldering, making it easy to make adjustments or reuse parts. Breadboards have a grid of holes with internal connections, so you can quickly assemble circuits by inserting components and connecting them with jumper wires. For more permanent projects, you might transition to **perforated boards (perfboards)** or **stripboards**, which allow soldered connections.

Wire Strippers and Cutters are essential for preparing wires to connect components. Stripping insulation carefully ensures a good electrical connection without damaging the wire. Having a good pair of cutters ensures you can trim wires neatly, preventing messy layouts that could lead to shorts.

Jumper Wires are used to connect components on a breadboard or perfboard. These come in various lengths and are color-coded for convenience, making it easy to trace connections in your prototype.

Measurement Tools: Checking Circuit Behavior

Designing a circuit is only half the battle—you also need to verify that it performs as expected. Measurement tools help you observe and analyze electrical behavior, allowing you to spot and fix problems.

A **Multimeter** is arguably the most important tool for circuit testing. It measures voltage, current, and resistance, making it invaluable for troubleshooting. For instance, if a component isn't working, you can use a multimeter to check whether it's receiving the correct voltage or if a connection is broken. Advanced multimeters also include features like continuity testing and diode checks, which are particularly useful for verifying circuit integrity.

An **Oscilloscope** goes beyond what a multimeter can do by displaying a waveform of a signal. This is especially useful for analyzing time-varying signals in circuits with transistors, capacitors, or ICs. With an oscilloscope, you can observe the shape, frequency, and amplitude of a signal, which helps you understand how your circuit processes input.

Logic Analyzers are specialized tools for digital circuits. They monitor multiple signals simultaneously, showing the timing and logic states (high or low) of digital signals. This is particularly useful for debugging communication protocols or verifying the functionality of microcontroller-based systems.

A **Power Supply with Adjustable Voltage and Current** is another key tool. It provides a reliable energy source for your circuit and allows you to control the voltage and current levels. This flexibility is crucial for testing circuits with components that have specific power requirements. Some power supplies include built-in meters to display the supplied voltage and current, adding an extra layer of convenience.

Assembly Tools: Putting Everything Together

Once you've designed your circuit and confirmed it works as a prototype, you'll want to assemble a more permanent version. For this, soldering tools and accessories are essential.

A **Soldering Iron** is used to join components and wires together on a circuit board. A soldering station with adjustable temperature is ideal, as it gives you precise control to avoid overheating components. Always use **lead-free solder** for safety and an **antistatic soldering mat** to protect sensitive parts.

Soldering Stands and Helping Hands hold your soldering iron and components steady, leaving your hands free to work more precisely. Helping hands often include magnifying glasses to help with small or delicate soldering jobs.

For cleaning up after soldering, a **Desoldering Pump (Solder Sucker)** or **Wick** can remove excess solder. These tools are also useful if you need to fix mistakes or replace components.

Simulation and Design Software

Before physically assembling your circuit, simulation tools can save you time and resources. These programs let you design and test circuits virtually, ensuring they work as intended before you start prototyping.

SPICE-Based Tools like LTSpice and Multisim allow you to simulate analog and digital circuits. You can model component behavior, test different configurations, and even measure voltages and currents—all in a virtual environment.

For designing printed circuit boards (PCBs), **CAD Software** such as KiCad, Eagle, or Altium Designer helps you lay out components and traces. These tools also generate files for manufacturing PCBs, streamlining the transition from prototype to finished product.

Specialized Tools for Advanced Projects

As you work on more complex circuits, you might need additional tools.

Function Generators produce test signals like sine waves, square waves, or pulses, which you can use to simulate inputs for your circuit. For example, if you're designing an amplifier, a function generator can provide a consistent test signal to evaluate its performance.

Thermal Cameras are useful for spotting overheating components in a circuit. Overheating can indicate issues like excessive current or poor ventilation, and a thermal camera provides a quick, non-contact way to identify these problems.

For working with microcontrollers or embedded systems, **Programmers and Debuggers** are indispensable. These tools upload code to microcontrollers and allow you to monitor their execution in real-time, making it easier to debug software and hardware interactions.

Organization and Safety

Good organization is as important as having the right tools. A well-organized workspace minimizes mistakes and makes it easier to find what you need. Use **storage bins or tackle boxes** to separate resistors, capacitors, and other small components. Label everything clearly, and keep your tools within easy reach.

Safety should also be a priority. Always wear **safety goggles** when soldering or cutting wires to protect your eyes from debris or accidental splashes. Use an **antistatic wrist strap** when handling sensitive components like microcontrollers or ICs to avoid damaging them with static electricity.

With the right tools and a thoughtful approach, designing and testing circuits becomes an enjoyable and rewarding process. As you build your toolkit, focus on quality and versatility, ensuring that your tools can grow with your skills and the complexity of your projects.

3.3 Building a Simple Circuit: Step-by-Step

Building your first circuit can feel like assembling a puzzle where every piece has a purpose. The process might seem daunting at first, but with a clear plan and the right tools, it becomes an enjoyable hands-on experience. In this section, we'll guide you through creating a simple LED circuit—a fundamental project for beginners that demonstrates how basic components work together.

Step 1: Plan the Circuit

Every project starts with a plan. Begin by identifying what you want the circuit to do. In this case, the goal is to create a circuit that lights up an LED when connected to a power source.

The circuit will include:

- A **power source**, like a 9V battery.
- An **LED**, which will light up.
- A **resistor**, to limit current and protect the LED.
- **Wires**, to connect the components.

Next, sketch a simple circuit diagram (schematic). The diagram should show the components and their connections. For this project, connect the positive terminal of the battery to the anode (longer leg) of the LED through a resistor. The cathode (shorter leg) of the LED connects back to the battery's negative terminal.

Step 2: Gather the Components and Tools

Once you have a clear plan, gather the necessary components and tools. You'll need:

- A 9V battery and a battery clip for easy connections.
- An LED (any standard red LED will work for this project).
- A resistor (use a 330-ohm resistor, calculated based on Ohm's Law; see **Chapter 1**).
- A breadboard for assembling the circuit without soldering.
- Jumper wires to connect the components on the breadboard.

Make sure you also have a multimeter on hand to test connections and measure voltages, as discussed in **3.2 Basic Tools for Circuit Design and Testing**.

Step 3: Set Up the Breadboard

Place the breadboard on your workspace. If you're unfamiliar with how a breadboard works, it's essentially a grid of holes connected internally to help you create temporary circuits.

Insert the components into the breadboard:

1. Place the LED on the breadboard. Insert the anode (longer leg) into one row and the cathode (shorter leg) into another row.
2. Insert one leg of the resistor into the same row as the LED's anode. The other leg can go into a different row.
3. Use a jumper wire to connect the row with the resistor to the positive rail of the breadboard.
4. Connect the cathode of the LED (its row) to the negative rail using another jumper wire.

Step 4: Connect the Power Source

Attach the battery to the battery clip and connect the clip's wires to the breadboard. The red wire connects to the positive rail, and the black wire connects to the negative rail. Always double-check the polarity, as reversing it can damage your LED.

Step 5: Test the Circuit

Before powering the circuit, inspect it to ensure all connections match your schematic. Use a multimeter to check continuity and confirm that the connections are secure. If everything looks good, connect the battery.

Once the power is connected, the LED should light up. If it doesn't, disconnect the battery immediately and troubleshoot.

Step 6: Troubleshoot Issues

If the LED doesn't light up, here are some common issues to check:

- **Polarity Errors:** Ensure the LED's anode and cathode are connected correctly. LEDs only work in one direction.
- **Loose Connections:** Verify that all wires and components are securely inserted into the breadboard.
- **Incorrect Resistor Value:** If the resistor value is too high, the current might be too weak to light the LED. Check the resistor's color bands or measure it with a multimeter.

Refer to **3.4 Common Issues and How to Troubleshoot Them** for more detailed troubleshooting tips.

Step 7: Modify and Experiment

Once your circuit works, try experimenting with it to learn more. Swap out the resistor for one with a different value and observe how it affects the LED's brightness. Add a second LED in parallel or series to see how the configuration changes the circuit.

You can also add a switch to the circuit. Place the switch in series with the resistor and LED to control when the circuit is closed or open. This small addition teaches you about controlling current flow manually.

Step 8: Document Your Work

Documenting your circuit is an important habit to develop. Take notes about the components you used, the schematic, and any modifications you made. Use software tools like Fritzing to create a clean digital representation of your breadboard layout and schematic.

Photos of your circuit and written observations will also be valuable if you revisit the project later or share it with others. Refer to **3.6 Tips for Organizing and Documenting Your Work** for best practices.

Step 9: Transition to a Permanent Build

If you're happy with the prototype and want to make it permanent, move it to a solderable perfboard or design a custom PCB. Use a soldering iron to secure the components and follow proper soldering techniques discussed in **3.5 Introduction to Breadboards and Soldering**.

Building Confidence Through Practice

By completing this simple circuit, you've taken a significant step in understanding how electrical components work together. As you gain confidence, you can tackle more complex projects involving transistors, microcontrollers, or integrated circuits. Each project builds your skills, helping you master the art of circuit design.

3.4 Common Issues and How to Troubleshoot Them

Even the simplest circuits can run into issues, from components not functioning as expected to complete failures. Troubleshooting is an essential skill for circuit design, allowing you to identify and resolve problems efficiently. In this section, we'll explore common issues you might face and provide strategies for diagnosing and fixing them.

1. Circuit Doesn't Work at All

One of the most frustrating problems is when nothing happens after powering your circuit.

- **Power Issues:** Check if the power source is correctly connected. As mentioned in **3.3 Building a Simple Circuit**, confirm the polarity of the connections—reversed connections can prevent the circuit from working. Verify that the power source (e.g., battery or power supply) is functioning and delivering the correct voltage using a multimeter.

- **Open Circuit:** An incomplete loop prevents current from flowing. Examine the circuit for loose wires, disconnected components, or faulty connections. If you're using a breadboard, ensure components are placed in the correct rows, as breadboards have specific internal layouts.

2. Components Not Functioning

Sometimes, individual components might fail to perform their roles.

- **Damaged Components:** Components like resistors, LEDs, and capacitors can be damaged during handling or by excessive voltage or current. Use a multimeter to test their functionality. For example, measure resistance to check a resistor or continuity to test an LED or diode.
- **Incorrect Orientation:** Some components, like LEDs, diodes, and electrolytic capacitors, are polarity-sensitive. Ensure their anode and cathode or positive and negative terminals are connected correctly. Reference **2.1 Resistors, Capacitors, and Inductors** for more details.

3. Unintended Short Circuits

A short circuit occurs when current bypasses the intended path, usually due to unintended contact between wires or components.

- **Check for Bare Wires:** Look for exposed wires that might be touching unintentionally. Insulate exposed areas using heat shrink tubing or electrical tape.
- **Inspect the Breadboard:** On breadboards, misplaced wires or components can easily cause connections between unintended rows. Carefully trace each connection against your schematic.
- **Use a Multimeter:** Switch your multimeter to continuity mode and check if there's a path where there shouldn't be one. A beeping sound indicates a short circuit.

4. Incorrect Voltage or Current Levels

Sometimes, components behave erratically because they're not receiving the correct voltage or current.

- **Voltage Drop:** Measure the voltage across components using a multimeter. If the voltage is lower than expected, it might indicate excessive resistance in the circuit or a weak power source.
- **Current Issues:** If a component doesn't perform as expected, check the current flowing through it. Use a multimeter in series with the component to measure the current and confirm it's within the expected range.

5. Overheating Components

Components like resistors, transistors, and ICs can overheat if they're overloaded.

- **Check Power Ratings:** Ensure components are rated to handle the power they're dissipating. For resistors, confirm the power is below their rated wattage using $P = I^2 \times R$.
- **Inspect for Shorts:** A short circuit can cause excessive current, leading to overheating. Refer to the section on short circuits for troubleshooting steps.
- **Add Cooling:** For components like transistors or voltage regulators, consider adding heatsinks or improving ventilation.

6. Noise or Unstable Signals

In circuits involving signals or communication, noise or instability can cause unexpected behavior.

- **Decoupling Capacitors:** Place capacitors near ICs or power supply lines to filter out high-frequency noise. See **2.2 Capacitors** for details on their filtering role.

- **Check Connections:** Loose or poor-quality connections can introduce noise. Ensure all solder joints or breadboard connections are secure.
- **Use Shielded Cables:** For sensitive circuits, shielded wires can prevent interference from external electromagnetic signals.

7. Misaligned or Faulty Schematic

A common mistake in circuit design is incorrectly interpreting or implementing a schematic.

- **Double-Check the Design:** Compare your physical circuit with the schematic. Look for discrepancies in component placement or wiring.
- **Simulate the Circuit:** If you're unsure about the design, use circuit simulation software to verify its behavior before assembling it physically.

8. Debugging Step-by-Step

When troubleshooting a circuit, it's essential to break the problem into smaller parts.

1. **Start with the Power Supply:** Confirm the power source is delivering the correct voltage and polarity.
2. **Test Individual Components:** Remove and test components one at a time. For example, if an LED isn't lighting up, test it separately with a resistor and power supply.
3. **Trace the Current Path:** Follow the flow of current through the circuit step-by-step, measuring voltages and checking continuity along the way.
4. **Isolate Sections:** If the circuit is complex, isolate and test smaller sections to identify the problem area.

When to Seek Help

If you've tried all the above steps and the circuit still doesn't work, don't hesitate to seek help. Share your schematic, component list, and observations with online forums or experienced individuals. Fresh eyes can often spot mistakes you might overlook.

Troubleshooting is as much about patience as it is about skill. Every issue you encounter teaches you something new about circuits, helping you build expertise over time. With practice, you'll develop the ability to diagnose and fix problems quickly, turning frustration into a rewarding learning experience.

3.5 Introduction to Breadboards and Soldering

Breadboards and soldering are two essential techniques for creating and working with circuits. Whether you're prototyping an idea or assembling a permanent design, these tools and methods provide the flexibility and durability needed for various stages of circuit development. Understanding when and how to use them is crucial for efficient circuit building and troubleshooting.

Breadboards: The Foundation of Prototyping

A breadboard is a reusable platform for assembling temporary circuits without the need for soldering. It's ideal for testing designs, experimenting with configurations, or learning the basics of circuit building.

Breadboards are made up of a grid of holes arranged in rows and columns. Beneath the surface, metal strips create connections between specific groups of holes. The layout is typically divided into two sections:

- **Terminal Strips:** These are the main working area where components and wires are inserted. Rows of holes in the terminal strips are internally connected, allowing

components in the same row to share electrical connections.
- **Power Rails:** Running along the top and bottom edges, these rails are used for distributing power. The positive and negative terminals of your power supply connect to these rails, simplifying connections for multiple components.

Using a Breadboard
Start by placing components and connecting them with jumper wires. For example, when assembling an LED circuit like the one described in **3.3 Building a Simple Circuit**, insert the LED, resistor, and wires into appropriate rows. Use the power rails for consistent voltage and ground connections.

Advantages of Breadboards

1. **Reusability:** Components can be added, removed, or repositioned without damage.
2. **Flexibility:** Breadboards support rapid design changes, making them ideal for prototyping and experimentation.
3. **Cost-Effective:** They eliminate the need for specialized equipment, such as soldering irons, during the initial design phase.

Limitations of Breadboards
While useful for prototyping, breadboards have limitations. They are unsuitable for high-frequency circuits due to noise and interference, and connections can be unreliable for heavy or vibration-prone setups.

Soldering: Making Circuits Permanent

Soldering is the process of joining electrical components to a permanent substrate, such as a perfboard or printed circuit board (PCB), using a metal alloy called solder. This technique is essential for creating durable and reliable circuits.

How Soldering Works

Soldering uses heat to melt solder, a mixture of tin and lead (or tin and other metals in lead-free varieties). The molten solder flows into the connection point, bonding the components and ensuring a good electrical connection. Once cooled, the solder hardens, creating a stable joint.

Basic Soldering Tools

To get started with soldering, you'll need:

- **Soldering Iron:** A tool that heats up to melt solder. Adjustable temperature models are recommended for better control.
- **Solder:** Choose lead-free solder for safety. Thin solder wires with a rosin core are ideal for electronics.
- **Sponge or Brass Tip Cleaner:** Keeps the soldering iron tip clean for precise work.
- **Helping Hands or Clamps:** Hold components steady, freeing your hands for soldering.
- **Desoldering Tools:** Include a desoldering pump or wick to correct mistakes or remove components.

Steps for Soldering a Simple Circuit

1. **Prepare the Workspace:** Work in a well-ventilated area to avoid inhaling fumes. Use a heat-resistant mat to protect your surface.
2. **Place Components on the Board:** Insert components into a perfboard or PCB, ensuring they are positioned according to the circuit design.
3. **Heat the Joint:** Touch the soldering iron tip to the point where the component lead meets the board's pad or hole.
4. **Apply Solder:** Feed the solder to the heated joint, allowing it to flow and create a shiny, smooth connection. Avoid applying solder directly to the iron.
5. **Inspect the Joint:** A good solder joint should be clean, shiny, and cone-shaped. Reheat and apply more solder if necessary.

6. **Trim Excess Leads:** Use wire cutters to remove any protruding component leads.

Tips for Effective Soldering

- Keep the soldering iron tip clean to ensure efficient heat transfer.
- Use the correct amount of solder—too little can create weak connections, while too much can cause shorts.
- Avoid overheating components, as excessive heat can damage sensitive parts like ICs.

Choosing Between Breadboards and Soldering

Breadboards and soldering serve different purposes in the circuit-building process. Use breadboards when prototyping, testing designs, or experimenting with new ideas. Once the circuit is finalized, soldering allows you to create a more durable and reliable version.

For example, after testing a circuit on a breadboard, you can transfer it to a perfboard for soldering, making it permanent. This transition reduces the risk of loose connections and is particularly important for circuits intended for long-term use.

Safety Considerations

Both breadboarding and soldering involve risks, so it's important to prioritize safety.

- **When Breadboarding:** Avoid powering the circuit while making adjustments to prevent short circuits or component damage.
- **When Soldering:** Wear safety goggles to protect your eyes from splashes of molten solder. Avoid touching the soldering iron's tip, as it can reach extremely high temperatures.

Building Circuits with Confidence

Breadboarding and soldering are complementary techniques that form the backbone of circuit design. Breadboards provide a quick and flexible way to experiment with ideas, while soldering ensures durability and long-term reliability. Mastering both techniques gives you the tools to prototype, refine, and finalize your designs, whether you're working on a simple project or a complex system. With practice, these skills become second nature, empowering you to bring your circuit ideas to life.

3.6 Tips for Organizing and Documenting Your Work

Good organization and thorough documentation are critical for successful circuit design and testing. Whether you're a beginner or an experienced engineer, maintaining a structured approach saves time, minimizes errors, and makes your projects easier to replicate or troubleshoot later. This section provides practical tips to help you organize your workspace, components, and project documentation effectively.

1. Organizing Your Workspace

A well-organized workspace allows you to focus on your project rather than wasting time searching for tools or components.

- **Dedicated Work Area:** Set up a specific area for working on circuits. Ideally, this space should have good lighting, a flat surface, and access to power outlets. A heat-resistant mat is useful for soldering tasks, as mentioned in **3.5 Introduction to Breadboards and Soldering**.
- **Tool Accessibility:** Arrange tools like multimeters, soldering irons, wire strippers, and pliers within easy reach. Use a pegboard or magnetic strip to hang frequently used tools, keeping them visible and accessible.
- **Component Storage:** Store components in labeled containers. Small drawers or tackle boxes are ideal for organizing resistors, capacitors, and other small parts.

Group similar components together and label each compartment with their values

2. Managing Components and Supplies

Working with circuits often involves numerous small components, and keeping them organized is essential for efficient work.

- **Label Everything:** Use clear labels for all component containers. For resistors and capacitors, include their values and tolerances. For ICs or transistors, add the part numbers or a brief description of their function.
- **Use Color-Coding:** Apply color-coded stickers or markers to differentiate components. For instance, you could assign colors to specific resistor values or capacitor types. This system makes it easier to find what you need at a glance.
- **Create a Checklist:** Maintain a checklist of the components and tools you'll need for each project. This ensures you have everything on hand before you begin, preventing interruptions.
- **Inventory Management:** Regularly check your supplies and reorder components as needed. Keeping a spreadsheet or using inventory management software can help track what you have and what needs replenishing.

3. Documenting Your Projects

Proper documentation is invaluable for recording your design process, troubleshooting issues, and sharing your work with others.

Schematic Diagrams

- Always start with a clear and accurate schematic. Software like KiCad, Fritzing, or Eagle makes it easy to create professional-looking schematics. Include all components, connections, and annotations to ensure anyone can replicate your circuit.

- Use standard symbols for components (e.g., resistors, diodes, transistors) to maintain clarity.

Circuit Layouts

- When prototyping on a breadboard, document the layout using diagrams or photos. Tools like Fritzing allow you to create digital breadboard layouts, which are especially helpful for sharing designs with others or revisiting projects later.
- Label the positions of key components and connections to avoid confusion if the circuit is disassembled or modified.

Test Results

- Record measurements and observations from testing. Note the expected vs. actual values for voltage, current, and other parameters. For example, if an LED isn't lighting up, document the measured voltage across it and any adjustments made to fix the issue.

Changes and Adjustments

- Keep track of any modifications made during the project. If you swapped a resistor for a different value or added a decoupling capacitor, document the change and why it was necessary. This helps you understand the evolution of your design.

4. Digital Tools for Documentation

Leveraging digital tools can streamline the documentation process and make your work more accessible.

- **Spreadsheet Software:** Use tools like Microsoft Excel or Google Sheets to maintain a list of components, track inventory, and record test results. Include columns for component type, value, quantity, and location in storage.

- **Project Management Apps:** Apps like Trello or Notion can help organize project timelines, tasks, and notes. Create a dedicated board for each project, with lists for design, testing, and troubleshooting.
- **Cloud Storage:** Save your schematics, layouts, and notes in cloud storage services like Google Drive or Dropbox. This ensures you can access your files from anywhere and provides a backup in case of data loss.
- **Simulation Software:** Tools like LTSpice or Multisim not only help design and test circuits but also allow you to save and share simulation files as part of your documentation.

5. Keeping a Project Logbook

A physical or digital logbook is an excellent way to maintain a chronological record of your projects.

- **Start with Objectives:** Begin each entry by stating the purpose of the project and what you hope to achieve.
- **Record Steps:** Document each step of the process, from initial design to final testing. Include details like component values, connections, and test results.
- **Note Challenges:** Write down any issues you encountered and how you resolved them. This helps you learn from mistakes and provides a reference for future projects.
- **Include Visuals:** Add sketches, photos, or screenshots of your circuit at different stages. Visuals make your logbook more informative and easier to understand.

6. Organizing Completed Projects

Once a project is finished, it's important to store the circuit and documentation properly for future use or reference.

- **Label Finished Circuits:** If you've soldered a permanent circuit, label it with the project name and date. For larger projects, include a brief description of its function.

- **Store Documentation Together:** Keep all related files, diagrams, and notes in a single folder or binder. If the project is digital, organize it into a well-structured directory on your computer or cloud storage.
- **Reuse Components When Possible:** If the project is no longer needed, salvage reusable components like resistors, capacitors, and ICs. Clean and store them for future use.

7. Sharing and Collaboration

Sharing your projects with others fosters learning and collaboration.

- **Online Platforms:** Publish your designs and findings on platforms like GitHub, Hackaday, or personal blogs. Include detailed documentation to make your work accessible and reproducible.
- **Community Forums:** Engage with online communities like Reddit's r/Electronics or the EEVblog forum to discuss your projects, seek advice, or provide feedback to others.
- **Workshops and Presentations:** If you enjoy teaching, consider presenting your projects at workshops or local maker spaces. Clear and well-documented projects make it easier to share your knowledge.

Final Thoughts

A well-organized approach to circuit design and thorough documentation of your work are invaluable habits that pay off in every project. By maintaining a tidy workspace, keeping track of your components, and recording your process in detail, you'll not only make your current projects smoother but also create a library of knowledge for future endeavors. These practices ensure that your ideas are accessible, replicable, and ready to share with the broader engineering community.

Chapter 4: Introduction to Electrical Systems

4.1 Power Generation and Distribution

Electricity powers nearly every aspect of modern life, from homes to industries. But before it reaches our devices, it undergoes a fascinating journey through generation and distribution systems. Understanding how power is created and delivered offers insight into the complex infrastructure that supports our daily lives.

Generating Electricity

Electricity generation begins at power plants, where various energy sources are converted into electrical energy. This process typically involves rotating machinery like turbines and generators, which use electromagnetic principles to produce electricity. The choice of energy source determines the type of power plant:

- **Thermal Power Plants:** These use fossil fuels like coal, natural gas, or oil. Heat energy from burning fuel converts water into steam, which drives a turbine connected to a generator.
- **Hydroelectric Power Plants:** Here, water from a reservoir flows through turbines, converting the potential energy of stored water into mechanical energy and then electricity.
- **Nuclear Power Plants:** These use controlled nuclear reactions to produce heat, which generates steam to drive turbines.
- **Renewable Energy Sources:** Wind turbines and solar farms generate electricity without burning fuels. Wind turbines harness kinetic energy from the wind, while solar panels use photovoltaic cells to convert sunlight directly into electricity.

Each method has its advantages and limitations, with factors like efficiency, environmental impact, and resource availability influencing the choice of generation method.

Transmission: High-Voltage Efficiency

Once generated, electricity needs to travel long distances to reach consumers. However, transmitting electricity at low voltages causes significant energy loss due to resistance in the transmission lines. To minimize these losses, electricity is stepped up to high voltages using transformers (see **4.3 Transformers and Their Role in Power Systems**).

High-voltage transmission lines, often supported by tall pylons, carry electricity across vast distances. The voltage levels can range from 110 kV to over 765 kV, depending on the system's design. While high-voltage transmission reduces energy loss, it introduces other challenges, such as insulation and safety concerns, which must be carefully managed.

Distribution: Bringing Power to Users

After transmission, electricity is stepped down to safer, usable voltage levels at substations. From here, distribution lines carry power to homes, businesses, and factories.

- **Primary Distribution:** High-voltage electricity (typically 11 kV or higher) is delivered to industrial users or urban substations.
- **Secondary Distribution:** Electricity is further reduced to standard household voltages, such as 120V or 230V, depending on the country.

The distribution network often includes:

- **Overhead Lines:** Common in rural areas due to lower costs.

- **Underground Cables:** More expensive but preferred in urban areas for safety and aesthetics.

Meters installed at consumers' premises measure energy usage, enabling utility companies to bill users accurately.

Grid Infrastructure and Stability

The electrical grid is a vast interconnected network of generation, transmission, and distribution systems. Grid stability is crucial to ensure a continuous supply of electricity. Imbalances between supply and demand can lead to blackouts or overloading. To maintain balance:

- **Load Balancing:** Utilities adjust the output of power plants or import/export electricity through interconnections with neighboring grids.
- **Smart Grids:** Advanced technologies allow real-time monitoring and automatic adjustments, improving efficiency and reliability.

Challenges in Power Systems

Modern power systems face various challenges, including:

- **Energy Losses:** Despite high-voltage transmission, energy losses in lines and transformers are inevitable.
- **Integration of Renewables:** Renewable energy sources like wind and solar are intermittent, requiring careful planning to maintain grid stability.
- **Aging Infrastructure:** Many grids were built decades ago and require upgrades to handle increasing demand and modern technologies.

Future of Power Generation and Distribution

The push for sustainable energy is driving innovation in power systems. Microgrids, which operate independently or in conjunction with the main grid, are gaining popularity for their ability to integrate local renewable energy sources. Smart grid technologies, battery storage systems, and decentralized energy generation are shaping the future of electricity distribution, making it more efficient and resilient.

Power generation and distribution systems form the backbone of our electrical infrastructure. They ensure that energy generated from diverse sources reaches users efficiently and reliably. Understanding these systems provides a deeper appreciation of the engineering marvels that keep our lights on and our devices running.

4.2 AC vs. DC Current: What You Need to Know

If you've ever plugged in a device or powered something with a battery, you've encountered alternating current (AC) and direct current (DC). These two types of electrical current are fundamental to the way we generate, distribute, and use electricity. While they serve the same purpose—delivering energy to power devices—they behave differently and are suited to different applications. Let's explore what sets them apart and why they're both essential in our modern world.

What is DC Current?

Direct current, or DC, is the simpler of the two. In a DC circuit, electricity flows in a single, constant direction, like water moving steadily through a straight pipe. This steady flow of electrons makes DC easy to understand and ideal for certain applications.

Batteries are the most common sources of DC power. When you connect a battery to a circuit, it creates a consistent voltage that pushes electrons in one direction. This is why devices like flashlights, smartphones, and laptops, which rely on batteries, use DC internally.

One of the main advantages of DC is its stability. Because the current doesn't change direction, it's perfect for sensitive electronics and devices requiring precise power levels. However, DC has its limitations. Over long distances, transmitting DC power becomes inefficient due to energy losses in the wires.

What is AC Current?

Alternating current, or AC, behaves differently. Instead of flowing in one direction, AC constantly changes direction, reversing many times per second. This back-and-forth motion is why it's called "alternating." The number of reversals per second is measured in hertz (Hz). For example, the standard in most of the world is 50 Hz, meaning the current alternates 50 times per second. In the United States, the standard is 60 Hz.

AC is the type of electricity that comes out of your wall sockets. It's generated by power plants and carried over long distances to homes and businesses. Unlike DC, AC can be easily transformed to higher or lower voltages using transformers (discussed in **4.3 Transformers and Their Role in Power Systems**). This flexibility makes AC ideal for power distribution.

The sinusoidal waveform of AC is another key feature. If you were to look at AC on an oscilloscope, you'd see a smooth wave-like pattern that rises and falls over time. This shape is not just a coincidence—it's the result of how AC is generated using rotating machines like turbines.

Key Differences Between AC and DC

To understand when and why AC or DC is used, let's break down their differences:

1. **Direction of Flow:**
    - DC flows steadily in one direction.
    - AC reverses direction periodically.
2. **Voltage Levels:**

- DC voltage is constant.
- AC voltage fluctuates, rising to a peak and falling to zero before reversing.

3. **Transmission Efficiency:**
   - AC is more efficient for long-distance transmission because its voltage can be stepped up or down using transformers.
   - DC suffers greater energy losses over long distances unless specialized high-voltage DC (HVDC) systems are used.

4. **Applications:**
   - DC is common in batteries, solar panels, and electronics.
   - AC powers homes, businesses, and large appliances like refrigerators and washing machines.

Why AC Became the Standard for Power Distribution

In the late 19th century, there was a famous "war of currents" between Thomas Edison, who championed DC, and Nikola Tesla, who promoted AC. Edison's DC systems were practical for short distances but struggled with efficiency over longer distances due to significant power losses. Tesla's AC system, however, solved this problem by using transformers to transmit power at high voltages, which reduced losses, and then stepping the voltage down for safe use.

Ultimately, Tesla's AC system won, and today it remains the backbone of power distribution worldwide. However, DC has made a resurgence in recent years, especially with technologies like solar power and battery storage.

Modern Applications of AC and DC

In today's world, AC and DC coexist, each serving specific roles:

AC Applications:

- **Power Grids:** As mentioned earlier, AC is used to transmit electricity over long distances and supply power to homes and businesses.
- **Large Appliances:** Refrigerators, air conditioners, and washing machines run on AC because it's the most practical form of power from the grid.

DC Applications:

- **Electronics:** Almost all modern electronic devices, from smartphones to televisions, use DC internally. That's why devices often include adapters or power supplies to convert AC from the wall into DC.
- **Renewable Energy Systems:** Solar panels and wind turbines generate DC power, which is stored in batteries or converted to AC using inverters for grid integration.

Hybrid Applications:

- **Electric Vehicles (EVs):** EVs run on DC because they use batteries, but charging stations often use AC, which must be converted to DC for the vehicle's battery.
- **HVDC Systems:** Some long-distance power transmission systems use high-voltage DC because it's more efficient for very long distances than traditional AC lines.

Making the Conversion: AC to DC and Vice Versa

Because the world relies on both AC and DC, devices known as converters bridge the gap:

- **Rectifiers:** Convert AC to DC. For example, a phone charger rectifies AC from the outlet to charge its DC battery.
- **Inverters:** Convert DC to AC. Solar inverters, for instance, take DC from solar panels and turn it into AC for use in homes or export to the grid.

These conversions allow the two types of current to complement each other rather than compete.

Why Understanding AC and DC Matters

For beginners in electrical engineering, grasping the differences between AC and DC is crucial because it impacts how circuits are designed and powered. When choosing components or designing systems, you need to know whether AC or DC is the better fit and how to convert between them if needed.

For example, if you're building a battery-powered device, you'll be working with DC. But if your circuit plugs into a wall socket, you'll need to account for AC and possibly include components to convert it to DC for sensitive parts. Understanding these fundamentals lays the groundwork for more advanced topics, such as power electronics and renewable energy systems.

The Balance Between Two Currents

While AC dominates power distribution, and DC reigns in portable and electronic devices, the two systems work together seamlessly in our modern electrical world. Thanks to innovations in conversion technologies, we can harness the strengths of both currents to power everything from a smartphone in your pocket to an entire city. Knowing the roles of AC and DC helps you appreciate the engineering behind the electricity we often take for granted.

4.3 Transformers and Their Role in Power Systems

Transformers are an indispensable part of modern power systems. These devices enable efficient transmission and distribution of electricity by altering voltage levels, making it possible to deliver power safely and effectively over long distances. Without transformers, the electricity we rely on daily would suffer from high losses, limited range, and unsafe voltage levels.

What is a Transformer?

A transformer is a static electrical device that transfers electrical energy between two or more circuits through electromagnetic induction. It consists of two windings, or coils, wrapped around a shared core:

- **Primary Winding:** Connected to the input power source, it generates a magnetic field.
- **Secondary Winding:** Induced by the magnetic field, it delivers the output power.

The number of turns in each winding determines whether the transformer increases (steps up) or decreases (steps down) the voltage.

How Transformers Work

Transformers operate based on **Faraday's Law of Electromagnetic Induction**, which states that a changing magnetic field in a coil of wire induces an electric current in a nearby coil.

When an alternating current (AC) flows through the primary winding, it creates a changing magnetic field in the core. This magnetic field induces a voltage in the secondary winding. The voltage transformation ratio depends on the ratio of turns in the primary and secondary windings:

$$\frac{V_1}{V_2} = \frac{N_1}{N_2}$$

Where V1 and V2 are the voltages across the primary and secondary windings, and N1 and N2 are the number of turns in each winding.

For example, if a transformer has 100 turns in the primary winding and 10 turns in the secondary, it steps down the voltage by a factor of 10. A 120V input would produce a 12V output.

Types of Transformers

Transformers are designed for specific applications and come in various forms:

1. **Step-Up Transformers:** Increase voltage from the primary to the secondary winding. These are commonly used in power plants to boost the voltage for long-distance transmission, reducing energy loss.
2. **Step-Down Transformers:** Decrease voltage from the primary to the secondary winding. These are used near homes and businesses to lower voltage to safe levels for appliances.
3. **Isolation Transformers:** Maintain the same voltage on both sides but isolate the primary and secondary circuits for safety or noise reduction.
4. **Autotransformers:** Use a single winding for both primary and secondary connections, offering a more compact and efficient design for certain applications.

Role of Transformers in Power Systems

Transformers play critical roles in the generation, transmission, and distribution of electricity.

1. Power Generation
Electricity is typically generated at moderate voltages, such as 11kV or 25kV. While these levels are sufficient for local use, they are too low for efficient long-distance transmission. Step-up transformers located at power plants boost the voltage to hundreds of kilovolts,

reducing current and minimizing resistive losses in transmission lines.

2. Power Transmission
High-voltage transmission lines carry electricity over vast distances. By stepping up the voltage, transformers reduce the current, which in turn decreases heat loss caused by resistance in the wires. Without this voltage increase, much of the generated power would dissipate as heat before reaching consumers.

3. Power Distribution
As electricity approaches its final destination, it must be stepped down to safer levels. Substations equipped with step-down transformers reduce the voltage to levels suitable for distribution networks (e.g., 11kV or 33kV). Local transformers, often seen on utility poles or near buildings, further reduce this to household voltages (e.g., 120V or 230V).

4. Industrial Applications
Industries often require customized voltage levels for machinery. Dedicated transformers supply power at specific voltages, ensuring compatibility and efficiency for industrial operations.

Efficiency and Losses in Transformers

Transformers are highly efficient, typically operating at efficiencies of 95–99%. However, some losses still occur:

- **Core Losses (Hysteresis and Eddy Currents):** Energy lost in the magnetic core due to alternating flux. Modern cores use laminated steel to minimize these losses.
- **Copper Losses:** Heat generated by resistance in the windings. Using high-quality conductive materials, like copper, reduces these losses.

Despite these losses, transformers remain the most efficient way to change voltage levels in AC systems.

Transformers and Renewable Energy

As renewable energy sources like wind and solar become more widespread, transformers are adapting to new demands. For example, wind turbines often generate electricity at low voltages, requiring step-up transformers to feed power into the grid. Similarly, solar farms use transformers to integrate DC output (converted to AC by inverters) into existing AC grids.

Challenges and Innovations

Modern power systems face several challenges:

- **Integration with Smart Grids:** Transformers now incorporate sensors and communication capabilities to monitor performance and optimize grid efficiency.
- **High-Voltage DC (HVDC) Systems:** While traditional transformers are designed for AC, HVDC systems use specialized transformers to handle the unique demands of DC transmission.
- **Compact Designs:** Innovations in materials and cooling systems are making transformers smaller and more efficient, addressing space and cost concerns in urban areas.

The Backbone of Power Systems

Transformers are the backbone of our electrical infrastructure, enabling the generation, transmission, and distribution of power at levels that are efficient and safe. From stepping up voltage for long-distance transmission to stepping it down for everyday use, transformers ensure that electricity can travel from power plants to your home with minimal losses. As the energy landscape evolves, these devices continue to play a vital role, adapting to new technologies and energy sources while maintaining the reliability we depend on.

4.4 Electric Motors and Generators

Electric motors and generators are foundational technologies in modern electrical systems. While they perform opposite functions—motors convert electrical energy into mechanical motion, and generators do the reverse—they are two sides of the same coin, sharing similar underlying principles. These devices have revolutionized industries, powering everything from household appliances to massive industrial equipment. Understanding how they work and their applications offers insight into the machinery that keeps our world moving.

The Basic Principle: Electromagnetic Induction

Both electric motors and generators operate on the principle of electromagnetic induction, discovered by Michael Faraday. This principle states that a changing magnetic field within a coil of wire induces an electric current.

- In a **generator**, mechanical motion (e.g., a spinning turbine) creates a changing magnetic field, inducing current and producing electricity.
- In a **motor**, an electric current in a magnetic field produces mechanical motion, spinning a shaft or rotor.

This interchangeability between electricity and motion makes motors and generators essential for energy conversion in power systems.

Electric Motors: Turning Electricity Into Motion

An electric motor is a machine that converts electrical energy into mechanical motion by exploiting the interaction between magnetic fields and current-carrying conductors.

How It Works
Electric motors typically consist of two main parts:

1. **Stator:** The stationary part of the motor, which generates a magnetic field.
2. **Rotor:** The rotating part, which interacts with the stator's magnetic field to produce motion.

When current flows through the windings in the motor, it creates an electromagnetic field. This field interacts with the magnetic field of the stator, generating a force that turns the rotor.

Types of Electric Motors

1. **DC Motors:** Powered by direct current, these motors rely on commutators and brushes to reverse the current's direction, ensuring continuous rotation. They are commonly used in toys, car starters, and some industrial equipment.
2. **AC Motors:** Powered by alternating current, these motors are simpler in design and are widely used in household appliances, fans, and pumps. AC motors can be further divided into:
    - **Induction Motors:** Use electromagnetic induction to generate motion.
    - **Synchronous Motors:** Rotate at a speed synchronized with the frequency of the AC supply.
3. **Brushless Motors:** Found in modern devices like drones and electric vehicles, these motors replace brushes with electronic controllers for higher efficiency and reliability.

Applications of Electric Motors

- **Household Appliances:** Washing machines, refrigerators, and vacuum cleaners rely on motors to perform their functions.
- **Transportation:** Electric vehicles, trains, and bicycles use motors to power their wheels.

- **Industrial Machinery:** Motors drive conveyor belts, lathes, and robotic arms, forming the backbone of manufacturing.

Generators: Creating Electricity From Motion

Generators perform the reverse function of motors, converting mechanical energy into electrical energy. They are central to power generation, turning turbines driven by water, steam, or wind into a source of electricity.

How It Works
Generators consist of:

1. **Rotor:** The rotating component, typically a coil or magnetic core.
2. **Stator:** The stationary part, usually containing coils of wire.

As the rotor spins, it cuts through the stator's magnetic field, inducing a voltage in the stator windings. This produces an alternating current (AC) or, in some cases, direct current (DC) depending on the generator's design.

Types of Generators

1. **AC Generators (Alternators):** Produce alternating current, commonly used in power plants and automotive systems.
2. **DC Generators:** Produce direct current, used in some niche applications where DC power is required.
3. **Portable Generators:** Provide temporary electricity during outages, often powered by internal combustion engines.
4. **Renewable Generators:** Found in wind turbines and hydroelectric plants, converting natural forces into electricity.

Applications of Generators

- **Power Plants:** Large-scale generators produce electricity for homes, businesses, and industries.
- **Backup Power:** Standby generators ensure critical systems remain operational during outages.
- **Renewable Energy Systems:** Wind and hydroelectric generators harness natural forces to supply clean energy.

Key Differences and Similarities

Though motors and generators perform opposite functions, their designs are remarkably similar. A motor can often function as a generator if operated in reverse, and vice versa. This dual functionality is why regenerative braking in electric vehicles works so effectively—motors temporarily act as generators to convert kinetic energy back into electrical energy for storage.

Differences in Operation

- Motors require an input of electrical energy, while generators produce electrical energy.
- Generators rely on external mechanical motion (e.g., a spinning turbine), while motors produce mechanical motion internally.

Similarities

- Both use electromagnetic induction as their fundamental principle.
- Both consist of rotors and stators to interact via magnetic fields.

Challenges and Advancements

Both motors and generators face challenges related to efficiency, reliability, and environmental impact. However, ongoing advancements are addressing these issues.

Efficiency Improvements: Modern designs, like brushless motors and superconducting generators, reduce energy losses and improve performance.

Integration With Renewables: Motors and generators are integral to renewable energy systems. For instance, wind turbines use generators to produce electricity, while electric vehicle motors contribute to greener transportation.

Compact Designs: Miniaturization has made motors and generators smaller and more efficient, enabling their use in portable devices like smartphones and drones.

Smart Technologies: Integration with IoT systems allows real-time monitoring and control, optimizing performance and reducing downtime in industrial settings.

The Power to Move and Create

Electric motors and generators are at the heart of our electrical and mechanical systems, driving innovation and enabling modern conveniences. Motors power everything from appliances to transportation, while generators ensure a steady supply of electricity for homes, businesses, and renewable energy systems. Understanding these devices not only reveals the brilliance of their design but also underscores their critical role in shaping the energy landscape of the future.

4.5 Basic Overview of Renewable Energy Systems

Renewable energy systems are transforming the way we generate and use electricity. Unlike traditional fossil fuels, which are finite and contribute to pollution, renewable energy relies on natural, sustainable resources such as sunlight, wind, and water. These systems have gained momentum in recent years, driven by the need to reduce carbon emissions, improve energy security, and create a more sustainable future. Let's dive into how these systems work, their components, and their growing importance in the global energy landscape.

What Makes Energy "Renewable"?

Renewable energy comes from sources that are naturally replenished. The sun rises each day, winds blow across the earth, and water flows in rivers and streams. These processes occur constantly, making their energy sources effectively infinite on a human timescale. This contrasts sharply with fossil fuels like coal and oil, which take millions of years to form and are being consumed far more quickly than they can be replenished.

Types of Renewable Energy Systems

There are several types of renewable energy systems, each suited to specific environments and applications.

1. Solar Power
Solar energy harnesses sunlight to generate electricity. There are two main types of solar systems:

- **Photovoltaic (PV) Systems:** These use solar panels made of semiconductor materials to convert sunlight directly into electricity. Each panel contains many solar cells, which generate DC power when exposed to light. This power is often converted to AC using inverters, as discussed in **4.2 AC vs. DC Current**.
- **Solar Thermal Systems:** Instead of generating electricity directly, these systems use sunlight to heat fluids. The heat can be used for domestic water heating or to generate steam that drives a turbine in large-scale solar power plants.

Solar panels are widely used in residential, commercial, and industrial settings. Their modular nature makes them ideal for anything from rooftop installations to massive solar farms.

2. Wind Power

Wind turbines capture the kinetic energy of the wind and convert it into electrical energy. When wind turns the turbine blades, they spin a rotor connected to a generator. The electricity produced is typically AC, making it easy to integrate into existing power grids.

Wind energy is most effective in areas with consistent wind patterns, such as coastal regions or open plains. Offshore wind farms are becoming increasingly popular due to stronger and more reliable winds over the ocean.

3. Hydropower

Hydropower uses the energy of moving water to generate electricity. The most common method involves dams, which store water in reservoirs. When released, the water flows through turbines, spinning them to produce electricity.

Small-scale systems, known as micro-hydropower, are also used to generate electricity in remote or rural areas with access to streams or rivers.

4. Geothermal Energy

This type of energy taps into the heat stored beneath the earth's surface. Geothermal power plants use this heat to produce steam, which drives turbines to generate electricity. Geothermal energy is especially useful in regions with significant volcanic activity, such as Iceland or parts of the United States.

5. Biomass

Biomass energy comes from organic materials such as wood, agricultural waste, or even algae. These materials are burned or processed to produce electricity or biofuels like ethanol and biodiesel. While biomass is renewable, it must be managed carefully to avoid environmental harm, such as deforestation.

How Renewable Energy Systems Work Together

No single renewable energy source can meet all energy demands. Instead, renewable energy systems often work in combination to maximize efficiency and reliability.

For example, a hybrid system might combine solar and wind power. During sunny days, solar panels generate electricity, while wind turbines take over when it's cloudy but windy. Adding battery storage allows excess energy generated during peak times to be stored and used later, ensuring a consistent supply.

Key Components of Renewable Energy Systems

While each type of system has unique components, they share some common features:

- **Energy Source:** The sun, wind, water, or geothermal heat serves as the input.
- **Conversion Device:** Solar panels, wind turbines, or other equipment convert the raw energy into electricity.
- **Inverters:** Convert DC power from sources like solar panels into AC power for use in homes and grids.
- **Storage Systems:** Batteries or other storage devices store excess energy for later use, crucial for addressing the intermittency of solar and wind power.
- **Grid Connection:** Many systems feed electricity directly into the grid, reducing dependence on traditional power plants.

Benefits of Renewable Energy Systems

1. **Environmental Sustainability:** Renewable energy systems produce little to no greenhouse gas emissions, helping combat climate change.
2. **Energy Independence:** By harnessing local resources like wind or sunlight, communities can reduce reliance on imported fuels.

3. **Economic Opportunities:** The renewable energy sector creates jobs in manufacturing, installation, and maintenance.
4. **Scalability:** Systems can be tailored to suit needs, from small rooftop solar panels to utility-scale wind farms.

Challenges of Renewable Energy

While renewable energy has many advantages, it also comes with challenges:

- **Intermittency:** Solar and wind energy depend on weather conditions, making energy storage crucial to balance supply and demand.
- **Upfront Costs:** While operational costs are low, the initial investment for equipment like solar panels or wind turbines can be high.
- **Land Use:** Large-scale installations, such as wind farms or solar plants, require significant land, which may conflict with other land uses.

The Future of Renewable Energy

The future of renewable energy looks promising. Advances in technology are making systems more efficient, affordable, and adaptable. For example, perovskite solar cells promise higher efficiency at a lower cost than traditional silicon-based panels. In wind power, larger turbines and floating offshore platforms are increasing output and expanding deployment options.

Moreover, energy storage technologies like lithium-ion and flow batteries are becoming more effective at addressing intermittency, enabling renewable energy to provide a consistent and reliable power supply.

Powering a Sustainable Future

Renewable energy systems are no longer just an alternative—they are becoming the cornerstone of the global energy transition. By harnessing the power of natural resources, these systems offer a cleaner, more sustainable way to meet our growing energy demands. While challenges remain, continued innovation and integration will ensure that renewable energy plays an ever-expanding role in shaping the future of power systems.

4.6 Everyday Applications of Electrical Systems

Electrical systems are so deeply embedded in our daily lives that we often take them for granted. From the moment we wake up and switch on the lights to the time we set an alarm before bed, electricity is there, silently powering our routines. But behind this convenience lies a vast network of systems and devices designed to meet our needs efficiently and safely. Let's explore how electrical systems touch different aspects of our lives, making the ordinary extraordinary.

Think about your morning routine. When you flip a light switch, an electrical circuit completes, illuminating the room. This simple act relies on an intricate network of power generation, transformers, and distribution lines (discussed in **4.1 Power Generation and Distribution**) to ensure electricity reaches your home. Appliances like coffee makers, toasters, and microwave ovens then spring to life, converting electrical energy into heat to prepare breakfast in minutes. Each of these devices relies on carefully designed electrical components and systems to perform their functions seamlessly.

Outside the home, transportation systems are another example of electrical systems in action. Electric trains and subways, for instance, rely on overhead lines or electrified tracks to power their motors, reducing reliance on fossil fuels. Even personal vehicles are shifting toward electric power, with electric cars gaining popularity for their efficiency and environmental benefits.

Charging stations scattered across cities are a testament to how electrical infrastructure is evolving to support this transition.

At work or school, electrical systems enable modern productivity. Computers, printers, and internet routers all depend on electricity to function. Behind the scenes, servers and data centers handle massive amounts of information, powered by complex electrical setups that ensure reliability and prevent downtime. Smart lighting and HVAC systems in buildings adapt to changing conditions, optimizing energy use and improving comfort.

Entertainment, too, is driven by electrical systems. Whether you're watching a movie on a television, playing a video game, or streaming music on a smart speaker, electricity powers the experience. Even wireless devices rely on electrical energy stored in batteries, which are charged using power supplied by the grid or renewable sources.

Healthcare is another critical area where electrical systems play a life-saving role. Hospitals depend on uninterrupted power to operate vital equipment like ventilators, imaging machines, and surgical tools. Backup generators ensure that even during power outages, these systems continue to function. At home, wearable health monitors and fitness trackers use electrical components to track vital signs, empowering individuals to take control of their health.

Streetlights and traffic signals demonstrate how electrical systems keep communities safe and organized. These seemingly simple systems prevent accidents, provide visibility at night, and guide traffic flow efficiently. Increasingly, smart technologies are being integrated into these systems, allowing remote monitoring and adaptive control to respond to real-time conditions.

Even leisure activities like camping have been transformed by portable electrical systems. Solar chargers, battery-powered lanterns, and electric stoves make outdoor experiences more convenient and comfortable, showing how far electrical technology has come in enhancing every aspect of our lives.

Everyday electrical systems may seem mundane, but they are the backbone of modern civilization. They bring light, comfort, and connectivity to our lives, making the impossible seem ordinary. By appreciating the systems behind these conveniences, we can better understand the immense engineering that powers our world.

Chapter 5: Introduction to Electronics

5.1 The Role of Electronics in Electrical Engineering

Electronics is at the heart of modern electrical engineering, bridging the gap between raw electrical power and the sophisticated systems that define today's technology. While electrical engineering focuses on generating, transmitting, and distributing power, electronics concentrates on controlling and processing that power to perform specific tasks. From the simplest LED circuit to advanced artificial intelligence systems, electronics transforms electrical energy into meaningful functionality.

At its core, electronics deals with components that control the flow of electrons. Unlike traditional electrical systems, which often involve high voltages and currents, electronic systems typically operate at lower power levels. This shift enables precision, miniaturization, and the ability to handle complex tasks like computation, communication, and automation.

The Evolution of Electronics

The rise of electronics has been transformative. In the early 20th century, vacuum tubes were the backbone of electronic systems, used for amplification and switching. These bulky components paved the way for the first radios, televisions, and even the earliest computers.

The invention of the transistor in 1947 marked a revolution. Transistors replaced vacuum tubes with a smaller, more efficient alternative, leading to the rapid development of consumer electronics. Integrated circuits (ICs), introduced in the 1960s, took this miniaturization further, packing thousands—and eventually billions—of transistors onto a single chip. Today, ICs power everything from smartphones to spacecraft.

Electronics in Modern Engineering

In electrical engineering, electronics plays several critical roles:

1. **Signal Processing:** Electronics is essential for manipulating electrical signals. For example, amplifiers boost weak signals in audio systems, while filters remove unwanted noise. Signal processing is fundamental in communication systems, radar, and medical devices like ECG machines.
2. **Control Systems:** Electronics enables precise control over electrical systems. For instance, microcontrollers regulate the speed of motors in industrial machinery, while sensors and actuators create feedback loops for automation.
3. **Data Handling:** Digital electronics, including processors and memory, form the backbone of modern computing. They allow electrical systems to process vast amounts of data for applications like weather prediction, financial modeling, and artificial intelligence.
4. **Energy Management:** Electronics optimizes energy usage in devices. Power electronics, a specialized branch, focuses on converting and controlling electrical power efficiently, such as in renewable energy systems and electric vehicles.

Impact on Daily Life

The influence of electronics extends far beyond industrial and scientific applications. Everyday devices like smartphones, televisions, and smart home systems are powered by electronic circuits that process inputs, make decisions, and produce outputs seamlessly. Wearable fitness trackers, for instance, rely on sensors and microprocessors to monitor health data and provide actionable insights.

The Future of Electronics in Electrical Engineering

As technology evolves, the role of electronics in electrical engineering continues to expand. Innovations like quantum computing, flexible electronics, and nanoscale devices promise to revolutionize industries. Electronics is also central to addressing global challenges, from creating energy-efficient systems to developing smarter cities and advancing healthcare technologies.

Understanding the role of electronics in electrical engineering is essential for anyone entering the field. It's the foundation for designing systems that are not only functional but also intelligent, efficient, and transformative.

5.2 Digital vs. Analog Signals

Signals are the language of electronics, representing information that devices can process, transmit, and interpret. These signals fall into two categories: analog and digital. Both are essential in electrical engineering, but they differ in how they convey information and where they are most effectively used. Understanding their distinctions, strengths, and applications is fundamental to working with modern electronic systems.

What Are Analog Signals?

Analog signals are continuous and vary smoothly over time. They can take on any value within a given range, much like how a dimmer switch smoothly adjusts a light's brightness. For instance, the electrical signal from a microphone converts sound waves into a continuous voltage that fluctuates in sync with the sound.

Because they are continuous, analog signals can represent complex, real-world phenomena like temperature, light intensity, or sound with high fidelity. However, this continuity also makes analog signals susceptible to degradation. Noise and interference can distort them, leading to inaccuracies, especially over long distances.

What Are Digital Signals?

In contrast, digital signals are discrete. They represent information in binary format—combinations of 0s and 1s. Instead of a smooth wave, a digital signal is like a series of steps, jumping between predefined levels. For example, a computer processes digital signals as high (1) or low (0) voltages to perform calculations or store data.

The key advantage of digital signals is their resilience to noise. Because they rely on distinct levels, small distortions or interference are less likely to cause errors. This makes digital systems more reliable for transmitting and processing data, particularly in noisy environments or over long distances.

Key Differences Between Analog and Digital Signals

While both types of signals serve essential purposes, their characteristics and applications differ significantly.

- **Continuity vs. Discreteness:** Analog signals are continuous, while digital signals are discrete.
- **Susceptibility to Noise:** Analog signals degrade easily with noise, whereas digital signals are more robust.
- **Precision and Complexity:** Analog signals can capture subtle variations, making them ideal for natural phenomena, while digital signals excel in precision and repeatability.
- **Storage and Transmission:** Digital signals are easier to store and transmit without loss of quality, which is why they dominate modern communication and data storage.

Real-World Examples

The choice between analog and digital signals depends on the application:

1. **Audio Systems:**
    - Analog: Vinyl records produce sound through a continuous groove that mirrors the original sound wave.
    - Digital: CDs and MP3s store sound as binary data, offering greater durability and easier replication.
2. **Television Broadcasting:**
    - Analog: Older TV systems used analog signals, which could suffer from snow or static interference.
    - Digital: Modern broadcasts use digital signals, delivering higher-quality images and sound with minimal interference.
3. **Sensors:**
    - Analog: Thermometers with mercury output a continuous measurement of temperature.
    - Digital: Digital thermometers convert the temperature reading into binary data for precise display and recording.

Why Digital Signals Dominate Modern Electronics

Digital signals are at the core of most modern electronics for several reasons:

- **Noise Immunity:** In digital systems, a voltage of 0.0V might represent a "0" and 5.0V might represent a "1." Even if noise alters the signal slightly to 0.2V or 4.8V, the system can still interpret it correctly. This robustness makes digital signals ideal for long-distance communication, such as in fiber optics or wireless networks.
- **Storage and Replication:** Digital data can be stored in memory devices like hard drives or flash storage without degradation over time. It can also be copied infinitely without loss of quality, unlike analog media such as tapes or records.
- **Processing Power:** Digital systems enable advanced computation. Microprocessors, which operate entirely with

digital signals, perform billions of calculations per second, powering everything from smartphones to supercomputers.

Where Analog Signals Still Shine

Despite the dominance of digital systems, analog signals remain irreplaceable in certain contexts:

- **Real-World Interfacing:** Most natural phenomena, like light, sound, and temperature, are inherently analog. Sensors often measure these in analog form before converting them into digital data.
- **High-Fidelity Applications:** In some audio and video systems, analog signals are preferred for their ability to capture subtle nuances, particularly in professional recording studios or high-end audio equipment.
- **Simplicity:** Analog circuits can be simpler and more cost-effective for basic tasks, such as amplifying a signal or turning on a light based on a threshold.

The Bridge Between Analog and Digital: Conversion

In many modern systems, analog and digital signals work together. Devices known as **Analog-to-Digital Converters (ADCs)** and **Digital-to-Analog Converters (DACs)** act as bridges between the two worlds:

- **ADCs:** Convert analog signals into digital data for processing or storage. For example, a digital camera uses an ADC to transform light (analog) into a digital image.
- **DACs:** Convert digital signals back into analog form, such as when a speaker plays audio from a digital file.

This interplay is essential in applications like smartphones, where the microphone captures analog sound, converts it to digital for processing, and then back to analog for playback.

A Dual Role in Electronics

Both analog and digital signals have vital roles in electronics. Analog signals capture the richness of the natural world, while digital signals provide precision and reliability for modern technology. Rather than competing, they complement each other, creating systems that leverage the strengths of both. Understanding these two types of signals is key to mastering electronic systems, as most designs require knowledge of when and how to use each effectively.

5.3 Basics of Logic Gates and Boolean Algebra

Logic gates and Boolean algebra are the foundation of digital electronics. These concepts form the basis of all modern computing and digital systems, from the microprocessors in your phone to the servers powering the internet. Understanding how logic gates work and how they relate to Boolean algebra is essential for anyone diving into the world of electronics.

What Are Logic Gates?

Logic gates are electronic circuits that process binary inputs (0s and 1s) and produce a single binary output. Each gate performs a basic logical function, such as AND, OR, or NOT. These gates are the building blocks of digital systems, enabling devices to perform tasks like computation, decision-making, and data processing.

Logic gates are implemented using transistors, as explained in **2.2 Diodes and Transistors**, but in practical use, they are often grouped into integrated circuits (ICs) for efficiency.

The Main Types of Logic Gates

Here's a quick overview of the most common types of logic gates and their functions:

1. **AND Gate**
    - **Function:** Produces an output of 1 only if all its inputs are 1.
    - **Truth Table:**

Input A	Input B	Output
0	0	0
0	1	0
1	0	0
1	1	1

2. **OR Gate**
    - **Function:** Produces an output of 1 if at least one input is 1.
    - **Truth Table:**

Input A	Input B	Output
0	0	0
0	1	1
1	0	1
1	1	1

3. **NOT Gate**
    - **Function:** Produces the opposite of its input.
    - **Truth Table:**

Input	Output
0	1
1	0

4. **NAND Gate**
    - **Function:** Produces the opposite of an AND gate's output.
    - **Truth Table:**

Input A	Input B	Output
0	0	1
0	1	1
1	0	1
1	1	0

5. **NOR Gate**
   - **Function:** Produces the opposite of an OR gate's output.
   - **Truth Table:**

Input A	Input B	Output
0	0	1
0	1	0
1	0	0
1	1	0

6. **XOR Gate (Exclusive OR)**
   - **Function:** Produces an output of 1 only if exactly one input is 1.
   - **Truth Table:**

Input A	Input B	Output
0	0	0
0	1	1
1	0	1
1	1	0

Boolean Algebra: The Language of Logic

Boolean algebra is a mathematical framework for representing and simplifying logical relationships. Developed by George Boole in the mid-19th century, it uses binary variables (1s and 0s) and logical operators to describe the behavior of logic gates and digital circuits.

The primary operators in Boolean algebra correspond directly to logic gates:

- **AND:** Represented as $A \cdot B$ or simply AB.
- **OR:** Represented as $A + B$.
- **NOT:** Represented as $\overline{A}$ (A bar).

Boolean Laws and Simplifications

Boolean algebra follows a set of rules and laws that help simplify expressions and optimize digital circuits. Some key laws include:

1. **Commutative Law:**
   - $A + B = B + A$
   - $AB = BA$

2. **Associative Law:**
   - $(A + B) + C = A + (B + C)$
   - $(AB)C = A(BC)$

3. **Distributive Law:**
   - $A(B + C) = AB + AC$

4. **Identity Law:**
   - $A + 0 = A$
   - $A \cdot 1 = A$

5. **Negation Law:**
   - $A + \overline{A} = 1$
   - $A \cdot \overline{A} = 0$

By applying these laws, complex expressions can be reduced to simpler forms, which in turn leads to more efficient circuit designs.

Practical Applications of Logic Gates and Boolean Algebra

1. **Arithmetic Circuits:**
   - Basic operations like addition are implemented using logic gates. For example, an XOR gate forms the basis of a half-adder, a fundamental building block of arithmetic circuits.
2. **Control Systems:**
   - Logic gates are used to create decision-making systems, such as elevator controllers or traffic lights.
3. **Memory Storage:**
   - Flip-flops, which store binary data, are constructed using combinations of NAND or NOR gates. These form the basis of memory in digital devices.
4. **Computing:**
   - Microprocessors rely on Boolean algebra and logic gates for everything from executing instructions to processing data.

Building the Digital World

Logic gates and Boolean algebra provide the foundation for all digital electronics. By combining these simple building blocks, engineers can design systems of incredible complexity, from smartphones to space probes. Mastering these basics is crucial for understanding how modern electronics work and for creating the digital systems that drive our future.

5.4 Introduction to Microcontrollers and Embedded Systems

Microcontrollers and embedded systems have revolutionized electronics by enabling intelligent control of devices, systems, and applications. These technologies lie at the core of modern automation, making everyday devices like washing machines, thermostats, and smartphones smarter and more efficient. Understanding how microcontrollers and embedded systems function is essential for anyone looking to design or work with intelligent electronic systems.

What is a Microcontroller?

A microcontroller (MCU) is a compact integrated circuit designed to perform specific tasks within a device or system. Unlike a general-purpose microprocessor found in computers, a microcontroller integrates multiple functionalities into a single chip, including:

- A **processor** (the brain of the system).
- **Memory** (to store programs and data).
- **Input/Output (I/O) peripherals** (to interact with external components like sensors or displays).

This self-contained nature makes microcontrollers ideal for controlling embedded systems, where cost, size, and power consumption are critical factors.

Microcontrollers operate based on a program stored in their memory. These programs, written in languages like C or Python, dictate how the microcontroller responds to inputs, processes data, and drives outputs.

What are Embedded Systems?

An embedded system is a dedicated computer system integrated into a larger device or application. It is designed to perform a specific function, often in real-time. Unlike general-purpose computers, embedded systems are task-specific and tightly coupled with the hardware they control.

Examples include:

- The control unit in a microwave oven that adjusts cooking times and power levels.
- The braking system in cars (ABS) that ensures safety by monitoring wheel speed and applying brakes efficiently.
- IoT devices, like smart thermostats, that combine sensors, communication modules, and microcontrollers to enhance automation.

Components of an Embedded System

An embedded system typically consists of:

1. **Microcontroller or Microprocessor:** Acts as the central control unit.
2. **Sensors:** Collect data from the environment (e.g., temperature, light, or motion).
3. **Actuators:** Execute actions based on processed data, such as turning on a motor or adjusting a valve.
4. **Memory:** Stores program code and runtime data.
5. **Communication Interfaces:** Facilitate data exchange with other devices, using protocols like UART, SPI, or I2C.

How Microcontrollers and Embedded Systems Work Together

The microcontroller serves as the core of an embedded system, orchestrating its functionality. For example, in a smart irrigation system:

- **Sensors** detect soil moisture levels.
- The **microcontroller** processes this data and determines whether watering is necessary.
- **Actuators**, like water pumps, are activated to irrigate the field.
- Communication modules send updates to the user via a smartphone app.

This integration of sensing, processing, and actuating is what makes embedded systems so powerful.

Common Microcontrollers and Their Applications

Microcontrollers come in a wide range of configurations to suit different applications.

1. **Arduino (AVR-based):**
    - Easy to program and widely used for hobbyist projects and prototyping.
    - Applications: DIY robotics, home automation, and educational tools.
2. **Raspberry Pi (Broadcom-based):**
    - A microcontroller and microprocessor hybrid for more complex tasks.
    - Applications: Media servers, IoT projects, and AI applications.
3. **STM32 (ARM-based):**
    - Offers high performance and advanced features for industrial and automotive applications.
    - Applications: Motor control, medical devices, and aerospace systems.
4. **ESP32 (Wi-Fi and Bluetooth enabled):**
    - Ideal for IoT projects that require wireless connectivity.
    - Applications: Smart home devices, wearables, and connected appliances.

Advantages of Using Microcontrollers in Embedded Systems

1. **Cost-Effective:** Microcontrollers are affordable and eliminate the need for separate processors and peripherals, reducing system costs.
2. **Low Power Consumption:** Many microcontrollers are designed to operate efficiently, making them ideal for battery-powered devices.

3. **Compact Size:** Their small form factor allows for integration into devices where space is limited.
4. **Flexibility:** A single microcontroller can be reprogrammed for different tasks, enhancing versatility during development.

Challenges of Embedded Systems

Despite their advantages, embedded systems face certain challenges:

- **Limited Resources:** Microcontrollers have constrained memory and processing power, requiring careful optimization of code and resources.
- **Real-Time Constraints:** Many embedded systems must respond instantly to inputs, necessitating efficient programming and reliable hardware.
- **Security Concerns:** As IoT devices proliferate, embedded systems become targets for cyberattacks. Ensuring secure communication and data storage is critical.

The Role of Software in Embedded Systems

Software plays a critical role in the functionality of embedded systems. Real-time operating systems (RTOS) are often used to manage tasks efficiently, ensuring timely responses in applications like automotive braking systems or medical monitors. For simpler systems, a loop-based programming approach might suffice, where the microcontroller continuously checks inputs and updates outputs.

Applications of Microcontrollers and Embedded Systems

Embedded systems are ubiquitous in today's world, with applications spanning industries:

1. **Consumer Electronics:** Smartphones, smartwatches, and gaming consoles rely on embedded systems for seamless functionality.
2. **Automotive Systems:** Modern cars feature dozens of microcontrollers for tasks like engine management, navigation, and safety features.
3. **Industrial Automation:** Factories use embedded systems for robotic arms, conveyor belts, and process monitoring to enhance productivity and precision.
4. **Healthcare Devices:** Heart rate monitors, insulin pumps, and imaging equipment depend on embedded systems for real-time operation and data accuracy.
5. **IoT Devices:** Smart home technologies like voice assistants, connected thermostats, and security cameras utilize embedded systems to integrate sensors, communication, and control.

Future Trends in Microcontrollers and Embedded Systems

As technology advances, microcontrollers and embedded systems are becoming more powerful, efficient, and integrated. Emerging trends include:

- **AI Integration:** Embedding artificial intelligence in devices to enable features like facial recognition, predictive maintenance, and autonomous decision-making.
- **Low-Power Systems:** Innovations like energy-harvesting microcontrollers are extending the battery life of IoT devices.
- **Edge Computing:** Performing data processing locally on embedded devices instead of relying on cloud services, reducing latency and enhancing security.

The Power Behind Intelligent Devices

Microcontrollers and embedded systems are the invisible workhorses of modern technology, enabling devices to sense, think, and act autonomously. By combining compact hardware with sophisticated software, they bring intelligence to systems across industries. Understanding these technologies opens the door to designing smarter, more efficient, and innovative solutions that shape the future of electronics and beyond.

5.5 Practical Applications of Electronics in Everyday Life

Electronics are so deeply woven into our daily lives that their presence often goes unnoticed. They power the devices we depend on, enhance our comfort, and drive advancements in virtually every field. From communication and entertainment to healthcare and transportation, the impact of electronics is profound. This section explores how electronics have shaped everyday life, highlighting key applications and their underlying technologies.

1. Communication

Modern communication systems are built on the foundation of electronics, allowing people to connect instantly across the globe.

Smartphones and Mobile Networks
Smartphones are a marvel of electronic engineering, combining microprocessors, memory, sensors, and communication modules in a compact device. These devices rely on wireless communication technologies like 4G, 5G, and Wi-Fi, all of which are powered by integrated circuits and radio-frequency electronics. Features like video calling, social media, and instant messaging wouldn't be possible without advancements in signal processing and antenna technology.

Television and Streaming
Televisions have evolved from analog cathode ray tube (CRT) devices to digital smart TVs. Modern TVs incorporate microcontrollers, image processors, and digital signal processors (DSPs) to deliver high-definition visuals and audio. Streaming platforms like Netflix and YouTube use servers powered by

advanced microprocessors to deliver content seamlessly over the internet.

2. Home Automation and Smart Living

Smart homes exemplify the integration of electronics and connectivity, enhancing convenience, security, and energy efficiency.

Smart Lighting and Thermostats

Devices like smart bulbs and thermostats use microcontrollers and sensors to adapt to user preferences. For instance, a thermostat can adjust the temperature based on the time of day or whether a room is occupied, optimizing energy consumption.

Home Security Systems

Modern security systems combine cameras, motion detectors, and alarms, all controlled by embedded systems. These systems can send alerts to your smartphone, allowing remote monitoring and control.

Voice Assistants

Smart speakers like Amazon Echo and Google Nest rely on advanced electronics to process voice commands and interact with other devices. They use microcontrollers, microphones, and neural processing units (NPUs) to enable natural language processing and smart home integration.

3. Entertainment and Media

Electronics have transformed how we experience entertainment, making it more interactive, immersive, and accessible.

Gaming Consoles and Virtual Reality

Gaming systems like the PlayStation and Xbox use high-performance processors, GPUs, and memory modules to deliver realistic graphics and gameplay. Virtual reality (VR) headsets add another layer of immersion by using motion sensors, accelerometers, and high-resolution displays.

Portable Music and Video
Portable devices like iPods, MP3 players, and tablets rely on digital storage and processing to deliver high-quality media on the go. Bluetooth headphones and wireless earbuds use low-power electronics to ensure seamless connectivity and extended battery life.

4. Healthcare and Medical Devices

The healthcare industry has been revolutionized by electronics, enabling accurate diagnostics, effective treatments, and improved patient monitoring.

Diagnostic Equipment
MRI scanners, CT machines, and X-ray systems use advanced electronics to capture detailed images of the human body. These devices rely on precise signal processing and control systems to operate effectively.

Wearable Health Monitors
Fitness trackers and smartwatches measure heart rate, steps, and sleep patterns using embedded systems and sensors. These devices provide real-time health data, empowering users to monitor their well-being proactively.

Life-Saving Devices
Pacemakers, insulin pumps, and defibrillators use microcontrollers and sensors to monitor and regulate critical bodily functions. These devices are designed to operate reliably and efficiently for extended periods.

5. Transportation

Electronics play a critical role in modern transportation, enhancing safety, efficiency, and convenience.

Electric and Autonomous Vehicles
Electric vehicles (EVs) rely on power electronics to manage their batteries and drive motors. Advanced driver-assistance systems (ADAS), like lane-keeping and adaptive cruise control, use sensors and microcontrollers to make driving safer. Fully autonomous vehicles integrate AI-powered electronics for decision-making and navigation.

Public Transportation
Systems like subways, trams, and buses use electronics for scheduling, ticketing, and real-time updates. Control systems in trains rely on embedded electronics to monitor speed, braking, and track alignment.

Aerospace and Aviation
Aircraft systems, from autopilot to in-flight entertainment, are powered by sophisticated electronics. Satellites use embedded systems to manage communication and navigation tasks, supporting everything from GPS services to weather monitoring.

6. Industry and Automation

Industrial automation relies heavily on electronics to optimize production, reduce costs, and ensure safety.

Robotics
Modern manufacturing uses robots controlled by embedded systems. These robots perform repetitive or dangerous tasks with high precision, improving efficiency and reducing human risk.

Sensors and Monitoring Systems
Factories use sensors to monitor variables like temperature, pressure, and flow rates. Microcontrollers process this data and adjust equipment in real-time to maintain optimal conditions.

Power Electronics
High-power devices like inverters and motor drives are essential

for industrial equipment. They use electronics to convert and control electrical energy, ensuring machines run efficiently.

7. Education and Learning

Electronics have transformed education, making learning more engaging and accessible.

Digital Classrooms
Smartboards, projectors, and e-learning platforms use electronics to create interactive educational experiences. Tablets and laptops have become essential tools for students, enabling online learning and collaboration.

DIY and Maker Movement
Platforms like Arduino and Raspberry Pi encourage hands-on learning of electronics. These devices allow students and hobbyists to build their own projects, fostering creativity and innovation.

8. Everyday Conveniences

Electronics also power the small conveniences we often overlook.

Kitchen Appliances
Devices like microwaves, coffee makers, and refrigerators use embedded systems for precise control. Smart appliances can connect to the internet, enabling remote monitoring and automation.

Personal Gadgets
From electric toothbrushes to hairdryers, electronics are present in countless personal care devices. These tools often include safety features like timers and temperature sensors to ensure optimal performance.

Portable Power Banks
Power banks use advanced battery management systems to store

and deliver energy efficiently, ensuring we can charge our devices anywhere.

9. Environmental Impact and Sustainability

Electronics are increasingly being used to promote sustainable living.

Renewable Energy Systems
Solar inverters, wind turbine controllers, and energy management systems use electronics to optimize power generation and distribution. These systems play a critical role in reducing dependence on fossil fuels, as discussed in **4.5 Basic Overview of Renewable Energy Systems**.

Smart Grids
Smart meters and energy-efficient devices help households and businesses monitor and reduce energy consumption. Electronics also enable grid operators to balance supply and demand, minimizing waste.

A World Transformed by Electronics

From the devices in our pockets to the systems that power entire cities, electronics shape every aspect of modern life. They enhance convenience, improve safety, and drive innovation in countless industries. As technology continues to evolve, electronics will play an even greater role in solving global challenges, from healthcare and transportation to sustainability and education. Understanding these applications not only highlights the importance of electronics but also inspires us to explore their potential for creating a better future.

Chapter 6: Careers and Further Learning in Electrical Engineering

6.1 Typical Career Paths for Electrical Engineers

Electrical engineering opens doors to a variety of rewarding careers, each offering unique opportunities to innovate, solve problems, and impact the world. Whether designing cutting-edge electronics or maintaining the infrastructure that keeps the lights on, electrical engineers are at the forefront of technology. Let's explore some of the most common paths an electrical engineer might take and what makes each role exciting and impactful.

1. Power Systems Engineer: Keeping the World Running
Power systems engineers are the unsung heroes of modern life. They design and manage the electrical grids that power homes, businesses, and industries. Their work is becoming increasingly complex as renewable energy sources like solar and wind are integrated into existing grids. Imagine designing systems that ensure a city's lights stay on even during storms or when demand peaks—this is the challenge power engineers tackle every day.

These engineers work on everything from high-voltage transmission lines to local substations, ensuring electricity is delivered safely and efficiently. With the push toward green energy, they're also at the forefront of building sustainable systems that reduce carbon emissions and reliance on fossil fuels.

2. Electronics Engineer: Building the Devices We Love
If you've ever marveled at the sleek design of a smartphone or the functionality of a modern television, you've experienced the work of an electronics engineer. These professionals design the circuits and components that make electronic devices possible. From prototyping new gadgets to refining their performance, electronics engineers are innovators.

In addition to consumer devices, they work on complex systems like medical imaging equipment and industrial controls. It's a field that combines creativity with precision, offering the satisfaction of seeing tangible results, whether it's a new wearable device or a high-performance speaker.

3. Embedded Systems Engineer: Bringing Intelligence to Machines

Embedded systems engineers are the brains behind devices that think. They design systems where software and hardware come together, enabling everyday items like washing machines, cars, and thermostats to perform smart functions. For example, the microcontroller in a smartwatch tracks your heart rate, processes the data, and syncs it with your phone—all thanks to embedded systems engineering.

The field offers a perfect mix of programming and electronics, making it ideal for problem solvers who love a hands-on approach. These engineers often work on futuristic projects, such as smart home devices and autonomous vehicles, creating solutions that improve how we live and interact with technology.

4. Telecommunications Engineer: Connecting the World

Telecommunications engineers make global communication possible. They design networks that allow billions of people to call, text, and stream content seamlessly. With the advent of 5G and fiber-optic networks, this field is constantly evolving, offering exciting opportunities to work with cutting-edge technologies.

Imagine being part of a team that ensures high-speed internet reaches remote areas or designing systems for satellite communications. Telecommunications engineers play a critical role in shaping how we stay connected, whether through mobile networks, satellite systems, or the internet.

5. Control Systems Engineer: Mastering Automation

In factories, airports, or even spacecraft, control systems engineers are the ones making sure everything runs smoothly. They design automated systems that monitor and adjust processes in real time, using sensors, actuators, and feedback loops.

For example, they might develop the control system for an assembly line that ensures products are built with precision, or design the braking system in a train that adjusts speed automatically based on conditions. This field blends electronics with physics and mathematics, offering a career that's both challenging and rewarding.

6. Robotics Engineer: Designing the Future of Machines

Robotics engineers take control systems a step further by creating machines capable of performing tasks autonomously. From robotic arms in manufacturing to drones used in agriculture, these engineers are at the cutting edge of innovation.

Robotics combines multiple disciplines, including electronics, mechanics, and artificial intelligence. It's an exciting field for those who dream of designing machines that can navigate and adapt to their environments. Whether it's building robots for space exploration or creating surgical robots for healthcare, robotics engineers are shaping the future.

7. Research and Development: Pushing the Boundaries

For those driven by curiosity, research and development (R&D) offers a chance to explore new frontiers in technology. Electrical engineers in R&D roles work in labs, experimenting with materials, devices, and systems that could redefine the industry.

They might develop next-generation batteries, work on quantum computing, or invent new ways to transmit power wirelessly. R&D roles are often found in academia, government agencies, or private tech companies, offering opportunities to collaborate with some of the brightest minds in the field.

8. Consulting and Entrepreneurship: Leading and Innovating

Some electrical engineers choose to leverage their expertise by becoming consultants or entrepreneurs. Consultants advise companies on optimizing systems, integrating new technologies, or solving technical challenges. Entrepreneurs, on the other hand, turn innovative ideas into businesses, creating new products or services that address unmet needs.

For example, an electrical engineer might start a company designing energy-efficient lighting systems or providing IoT solutions for smart cities. These paths offer creative freedom and the potential to make a significant impact.

Electrical engineering is not just a career—it's a journey filled with opportunities to shape the world. Whether you're drawn to the technical challenge of designing circuits, the satisfaction of keeping power systems running, or the excitement of working on futuristic projects, there's a path for you. The field is constantly evolving, promising a lifetime of learning and innovation. The question is not whether there's a place for you in electrical engineering—it's which path you'll choose to make your mark.

6.2 Skills and Certifications to Advance Your Career

Success in electrical engineering requires more than a strong foundation in theory—it demands a diverse skill set and ongoing professional development. As technologies evolve, acquiring new skills and certifications becomes crucial to staying competitive and opening doors to advanced opportunities. This section highlights key skills and certifications that can propel your career in electrical engineering.

1. Essential Technical Skills

Circuit Design and Analysis
A strong understanding of circuit design is fundamental to electrical engineering. Beyond the basics, mastering tools like SPICE simulation software and PCB design platforms such as KiCad or Altium Designer can greatly enhance your ability to create efficient and reliable systems.

Programming and Embedded Systems
As discussed in **5.4 Introduction to Microcontrollers and**

Embedded Systems, programming is an increasingly vital skill. Languages like C, Python, and MATLAB are widely used in embedded systems, automation, and signal processing. Familiarity with microcontroller platforms like Arduino or STM32 can give you an edge in projects involving IoT or robotics.

Power Systems and Renewable Energy

With the growing focus on sustainable energy, expertise in power systems, energy storage, and renewable technologies is highly sought after. Understanding power electronics and grid integration can open opportunities in the rapidly expanding green energy sector.

Signal Processing and Communication

For engineers working in telecommunications or electronics, skills in signal processing and modulation techniques are critical. Knowledge of tools like MATLAB and LabVIEW, coupled with an understanding of protocols such as UART, SPI, and I2C, is invaluable.

Automation and Control Systems

Control systems engineers benefit from expertise in PID control, SCADA systems, and industrial automation. Learning how to program programmable logic controllers (PLCs) and work with industrial robots is especially useful for manufacturing and aerospace applications.

2. Soft Skills for Success

Problem-Solving and Critical Thinking

Electrical engineering often involves troubleshooting and optimizing systems. Being able to approach problems methodically, analyze data, and propose innovative solutions is a key skill that sets exceptional engineers apart.

Communication and Teamwork

Whether working in R&D, consulting, or project management, clear communication is vital. Engineers frequently collaborate with multidisciplinary teams and need to present ideas, document progress, and explain technical concepts to non-technical stakeholders effectively.

Time Management and Organization
Balancing multiple projects, meeting deadlines, and maintaining clear documentation are essential in any engineering role. Developing a disciplined approach to organizing tasks and priorities ensures efficiency and reliability.

3. Key Certifications to Boost Your Career

Certifications validate your expertise and demonstrate your commitment to professional growth. Here are some of the most recognized certifications for electrical engineers:

Professional Engineer (PE) License
This certification is essential for engineers who want to offer their services directly to the public. It requires passing the Fundamentals of Engineering (FE) exam, gaining relevant work experience, and passing the PE exam. A PE license enhances your credibility and is often required for leadership roles in engineering projects.

Certified Engineering Technician (CET)
For those focusing on technical roles, the CET credential demonstrates practical expertise in implementing and maintaining systems.

Certified LabVIEW Developer (CLD)
LabVIEW is widely used in test, measurement, and control systems. Becoming a Certified LabVIEW Developer validates your ability to create reliable and efficient applications using this platform.

LEED Certification
Engineers working in sustainable building design or energy systems can benefit from Leadership in Energy and Environmental

Design (LEED) certification. It demonstrates proficiency in energy-efficient and environmentally friendly practices.

IoT and Embedded Systems Certifications
Certifications from organizations like ARM or vendors like Cisco validate expertise in embedded systems, IoT platforms, and connected devices. Examples include:

- ARM Accredited Engineer (AAE).
- Cisco Certified Internet of Things Specialist.

Power Systems Certifications
For those in energy and power distribution, certifications like:

- **IEEE Power Systems Certification**: Focuses on advanced power systems analysis and design.
- **Grid Edge Certified Energy Engineer**: Highlights expertise in modern energy systems, including renewable energy integration.

Project Management Certifications
Engineers aspiring to management roles can benefit from project management certifications such as:

- Project Management Professional (PMP).
- Certified ScrumMaster (CSM) for agile environments.

4. Staying Current with Emerging Technologies

Electrical engineering is a field in constant evolution. Staying relevant requires a proactive approach to learning about emerging technologies, such as:

- **Artificial Intelligence (AI):** Understanding how AI integrates with electronics can give you a competitive edge, particularly in automation and robotics.

- **Quantum Computing:** While still in its infancy, gaining familiarity with quantum technologies can prepare you for cutting-edge roles.
- **5G and Beyond:** Engineers in telecommunications should stay informed about advancements in wireless communication and network optimization.

5. How to Choose the Right Skills and Certifications

Not every skill or certification is relevant to every career path. For example:

- If you're interested in renewable energy, focus on power systems and LEED certification.
- If your goal is a career in automation, prioritize learning PLC programming and obtaining certifications in robotics.
- For those pursuing leadership roles, project management certifications like PMP can complement your technical expertise.

Assess your interests and career goals, then align your skill development and certifications to match.

A Lifelong Commitment to Growth

The field of electrical engineering rewards those who invest in their skills and knowledge. By mastering technical and soft skills, earning certifications, and staying attuned to emerging technologies, you position yourself for success in an ever-changing landscape. The path to advancement is not a single step but an ongoing journey, and every skill you acquire builds the foundation for your next opportunity.

6.3 Resources for Lifelong Learning (Books, Online Courses, Communities)

Electrical engineering is a dynamic field, with new technologies and methodologies constantly emerging. To stay competitive and innovative, engineers must commit to lifelong learning. Whether you're a seasoned professional or just starting out, having access to high-quality resources can help you expand your knowledge, refine your skills, and connect with others in the field. This section provides an overview of some of the best books, online courses, and communities that support continuous learning in electrical engineering.

Books: The Timeless Source of Knowledge

Books remain one of the most reliable ways to deepen your understanding of core concepts and explore advanced topics. Here are some highly recommended texts for electrical engineers:

1. For Foundational Knowledge

- *The Art of Electronics* by Paul Horowitz and Winfield Hill: Often regarded as the bible of electronics, this book provides a practical guide to circuit design and analysis, from basic components to advanced applications.
- *Electrical Engineering 101* by Darren Ashby: A beginner-friendly introduction that covers fundamental principles with humor and clarity, perfect for refreshing core concepts.

2. For Specialized Topics

- *Power System Analysis and Design* by J. Duncan Glover and Mulukutla S. Sarma: Ideal for those interested in power systems, this book delves into the intricacies of generation, transmission, and distribution.
- *Microelectronic Circuits* by Adel S. Sedra and Kenneth C. Smith: A detailed exploration of circuit design and microelectronics, often used as a reference in academia and industry.

3. For Emerging Fields

- *Learning the Art of Electronics: A Hands-On Lab Course* by Thomas C. Hayes and Paul Horowitz: A practical companion to *The Art of Electronics*, focusing on hands-on experimentation.
- *Introduction to Embedded Systems* by Edward A. Lee and Sanjit A. Seshia: Essential for understanding how to design and implement embedded systems, as discussed in **5.4 Introduction to Microcontrollers and Embedded Systems**.

Online Courses: Learn Anywhere, Anytime

Online learning platforms have made it easier than ever to access high-quality education from experts worldwide. Here are some of the best platforms and courses for electrical engineers:

1. General Platforms

- **Coursera:** Offers university-level courses from institutions like Stanford and MIT. Recommended courses include:
  - *Introduction to Electronics* by Georgia Institute of Technology.
  - *Power Electronics* by University of Colorado Boulder.
- **edX:** Features programs like the MicroMasters in Power Engineering from Delft University of Technology.

2. Practical and Hands-On Learning

- **Udemy:** Known for its affordability and diverse offerings, including:
  - *Mastering Solar Energy* for renewable energy enthusiasts.
  - *Arduino Step-by-Step* for embedded systems.
- **Pluralsight:** Focuses on tech skills, including circuit design, programming, and IoT.

3. Advanced Specializations

- **MIT OpenCourseWare:** Free courses on advanced topics like control systems, electromagnetics, and machine learning for engineers.
- **IEEE Learning Network:** Offers specialized training in areas like 5G, smart grids, and renewable energy.

Communities: Learn and Grow Together

Connecting with peers, mentors, and industry experts is an invaluable part of lifelong learning. Being part of a community allows you to share knowledge, seek advice, and stay motivated.

1. Online Forums and Communities

- **Reddit:** Subreddits like r/ElectricalEngineering and r/Electronics are excellent places to discuss concepts, ask questions, and share projects.
- **EEVblog Forum:** A popular platform for discussing electronics design, troubleshooting, and tools.
- **All About Circuits:** Offers a mix of articles, forums, and tutorials, fostering a vibrant community of engineers and enthusiasts.

2. Professional Organizations

- **IEEE (Institute of Electrical and Electronics Engineers):** A global organization that provides access to conferences, journals, and local chapters. Membership includes valuable resources like IEEE Xplore and networking opportunities.
- **IET (Institution of Engineering and Technology):** Offers professional development programs and connects engineers through local and international events.

3. **Maker Communities**

- **Hackaday.io:** A hub for engineers and makers to share projects, collaborate, and gain inspiration.
- **Maker Faire Events:** Encourage creativity and innovation, showcasing projects in robotics, IoT, and renewable energy.

Podcasts and YouTube Channels

For bite-sized learning and inspiration, podcasts and video content are excellent options:

1. Podcasts

- *The Amp Hour:* A podcast about electronics design, engineering culture, and industry trends.
- *MacroFab Engineering Podcast:* Covers topics ranging from prototyping to manufacturing in the electronics world.

2. YouTube Channels

- **EEVblog:** Hosted by Dave Jones, this channel provides deep dives into electronics design, troubleshooting, and industry insights.
- **GreatScott!:** Focuses on DIY electronics projects and tutorials.
- **Adafruit Industries:** Offers videos on building projects using microcontrollers like Arduino and Raspberry Pi.

Conferences and Workshops

Attending industry events provides a unique opportunity to learn about cutting-edge developments and network with professionals. Some notable events include:

- **IEEE International Conference on Electronics and Communication:** Focused on academic and industrial advancements in the field.
- **Maker Faire:** A hands-on event showcasing innovations in electronics, robotics, and IoT.
- **Greenbuild International Conference and Expo:** Explores sustainable design and energy-efficient technologies.

Workshops hosted by local makerspaces or university extension programs can also provide hands-on experience in topics like circuit design, PCB layout, or embedded systems programming.

Lifelong Learning Strategies

To make the most of these resources, consider adopting the following strategies:

- **Create a Learning Plan:** Identify areas you want to develop, whether it's mastering a new tool, exploring an emerging technology, or earning a certification.
- **Dedicate Time Weekly:** Consistency is key. Set aside time each week to read, watch tutorials, or work on a project.
- **Join Study Groups:** Collaborating with peers can help reinforce concepts and keep you motivated.
- **Document Your Progress:** Maintaining a portfolio of completed courses, certifications, and projects can demonstrate your expertise to potential employers.

Embrace the Journey

Lifelong learning is not just a necessity in electrical engineering—it's a rewarding journey. By leveraging books, online courses, communities, and events, you can stay ahead in this ever-evolving field. These resources are not only tools for acquiring knowledge but also gateways to discovering new interests, solving complex challenges, and making meaningful contributions to the world of technology.

6.4 Tips for Success in Electrical Engineering

A successful career in electrical engineering requires more than technical expertise. It demands a combination of problem-solving skills, adaptability, and a commitment to continuous learning. Whether you're a student, a new graduate, or an experienced professional, these tips can help you thrive in this dynamic and rewarding field.

1. Build a Strong Foundation

Master the fundamentals of electrical engineering early in your journey. Topics like circuit design, electromagnetics, and control systems form the backbone of the field. As discussed in earlier chapters, skills in areas such as programming and embedded systems are increasingly important in modern engineering. Focus on gaining hands-on experience with tools and platforms like Arduino, SPICE, and MATLAB to reinforce theoretical concepts.

2. Embrace Lifelong Learning

The pace of innovation in electrical engineering means there's always something new to learn. Stay updated on emerging technologies such as renewable energy systems, IoT, and AI integration. Utilize resources like online courses, professional certifications, and industry publications to keep your knowledge relevant, as highlighted in **6.3 Resources for Lifelong Learning**.

3. Cultivate Problem-Solving and Critical Thinking Skills

Electrical engineers often face complex challenges, from debugging circuits to optimizing systems. Approach problems methodically: break them into smaller parts, analyze potential solutions, and test thoroughly. Developing these skills will help you tackle everything from troubleshooting hardware issues to designing innovative products.

4. Network and Seek Mentorship

Connecting with peers and industry professionals can open doors to new opportunities and provide valuable insights. Join professional organizations like IEEE or local engineering groups to meet mentors and collaborators. Attending conferences, workshops, and meetups can also expand your professional network and inspire new ideas.

5. Develop Soft Skills

Technical skills are essential, but communication, teamwork, and project management are equally important. Engineers often work in multidisciplinary teams, so being able to convey ideas clearly and collaborate effectively is vital. Strong documentation habits and the ability to present complex concepts to non-technical stakeholders are invaluable assets.

6. Stay Resilient and Adaptable

The engineering field is constantly evolving, and setbacks are inevitable. Embrace challenges as learning opportunities, and stay adaptable to new technologies and approaches. A positive mindset and a willingness to pivot when needed will set you apart.

Chapter 7: Practical Projects in Electrical Engineering

Before diving into the exciting projects in this chapter, it's time to roll up your sleeves and bring theory to life. Everything you've learned so far—from the basics of circuits and components to advanced systems and concepts—has prepared you for this moment. Hands-on projects are where the magic happens. They allow you to transform abstract ideas into tangible results, sparking creativity and deepening your understanding. Let's embark on this journey of discovery and creation—it's as rewarding as it is fascinating!

7.1 Why Hands-On Experience is Crucial

Electrical engineering is a field that thrives on the interplay between theory and practice. While understanding principles like Ohm's Law or the function of transistors is essential, the true mastery of electrical engineering comes from applying these concepts in real-world scenarios. Hands-on experience bridges the gap between knowledge and execution, offering invaluable lessons that no textbook or lecture can provide.

Turning Knowledge into Understanding

When working on a practical project, theoretical concepts come alive. For example, a formula like V=IR feels abstract until you build a circuit and see firsthand how varying the resistance affects voltage and current. Similarly, understanding how a diode blocks reverse current becomes far more intuitive when you wire it into a circuit and test it with different polarities.

Hands-on experience solidifies these connections, turning passive learning into active problem-solving. It teaches not just the "what" and "why," but also the "how" of electrical engineering. This practical approach builds a foundation of understanding that you'll carry throughout your career.

Building Problem-Solving Skills

In the real world, things rarely work perfectly on the first try. Wires might come loose, components could be faulty, or unexpected interactions may disrupt your circuit. These challenges are where hands-on work shines.

Troubleshooting a circuit teaches you to think critically and approach problems methodically. You learn to isolate variables, test hypotheses, and refine solutions—skills that are invaluable in any engineering discipline. Beyond technical problem-solving, you develop patience and resilience, qualities that help you persevere through complex projects.

Igniting Creativity

Practical projects give you the freedom to experiment. When designing a system, you have the opportunity to explore different approaches, try new components, and push the limits of what's possible.

For instance, adding a sensor to a simple circuit might spark ideas for automation, or integrating a microcontroller could inspire a completely new project. These moments of creative exploration often lead to innovative designs and a deeper appreciation of the tools at your disposal.

Building Confidence Through Tangible Results

There's nothing quite like the satisfaction of seeing your project come to life. Whether it's an LED lighting up, a motor spinning, or a sensor providing accurate readings, these small victories build confidence in your abilities. Each success reinforces your skills and motivates you to take on more ambitious challenges.

Moreover, hands-on experience gives you the confidence to step into professional environments, where practical skills and troubleshooting abilities are highly valued. Employers and colleagues will recognize your capability to transform ideas into functioning systems, setting you apart as a competent and reliable engineer.

A Path to Lifelong Learning

As you engage in hands-on projects, you'll encounter new technologies, tools, and methods. These experiences naturally lead to continuous learning, whether it's mastering a new software platform, exploring advanced sensors, or experimenting with renewable energy systems. This mindset of curiosity and exploration ensures that you stay adaptable and relevant in a rapidly evolving field.

Practical experience is the heart of electrical engineering. It transforms theoretical knowledge into actionable skills, nurtures creativity, and builds problem-solving expertise. As you dive into the projects in this chapter, remember that each challenge you face is an opportunity to grow. The journey of making, testing, and refining your designs will not only deepen your understanding but also ignite your passion for creating solutions that work in the real world. Let's get started!

7.2 Creating a Dynamic LED Lighting System

Building a dynamic LED lighting system is a perfect project to explore how electronics can interact with the environment. By integrating sensors like photoresistors (light sensors) or PIR (Passive Infrared) sensors for motion detection, you'll create a system where LEDs respond to specific conditions. This project introduces you to condition-based control—a foundational concept in automation and smart devices. Let's dive in!

The Concept: Responsive Lighting

Unlike a simple LED circuit that lights up when powered (as discussed in **3.3 Building a Simple Circuit**), this system reacts to its surroundings. For example:

- **Light-Activated LEDs:** The LED turns on when it gets dark, useful for nightlights or garden path lighting.
- **Motion-Activated LEDs:** The LED lights up when it detects movement, commonly used in security systems or automatic hallway lighting.

This project combines sensing, logic, and actuation, showcasing how individual components work together to create intelligent behavior.

Components and Tools Needed

To build your dynamic LED lighting system, you'll need:

- **Core Components:**
    - 1x Photoresistor or PIR sensor (depending on your chosen application).
    - 1x LED (or multiple for a more complex system).
    - 1x Resistor (appropriate for your LED, typically 220Ω).
    - 1x NPN transistor (e.g., 2N2222) for amplifying the signal.
    - Optional: 1x Arduino for advanced control and logic (if desired).
- **Power Source:** Battery pack or 5V DC adapter.
- **Breadboard and Jumper Wires:** For assembling the circuit.
- **Multimeter:** For testing connections and troubleshooting.

Step 1: Understanding Your Sensor

1. **Photoresistor:**
    - A photoresistor changes its resistance based on light intensity. When it's dark, its resistance increases, and when it's bright, resistance decreases.
    - By creating a voltage divider with the photoresistor and a fixed resistor, you can generate a variable voltage signal based on light levels.

2. **PIR Sensor:**
    - A PIR sensor detects infrared radiation emitted by warm objects, such as humans or animals. When movement is detected, the sensor outputs a high signal (usually 3.3V or 5V), which can directly trigger your circuit or microcontroller.

Step 2: Designing the Circuit

1. **For a Photoresistor-Based System:**
    - Create a voltage divider with the photoresistor and a fixed resistor. Connect the junction between them to the base of an NPN transistor.
    - The emitter of the transistor goes to ground, and the collector connects to the LED and its current-limiting resistor.
    - When the photoresistor detects low light, the increased base voltage turns on the transistor, allowing current to flow through the LED.
2. **For a PIR Sensor-Based System:**
    - Connect the output of the PIR sensor to the base of the transistor through a small resistor (1kΩ).
    - As with the photoresistor system, the LED and its resistor connect to the transistor's collector, and the emitter is grounded.
    - When motion is detected, the PIR outputs a high signal, turning on the transistor and lighting the LED.

Step 3: Optional Advanced Control with Arduino

If you want more control over your lighting system, integrate an Arduino microcontroller:

- **Setup:** Connect the sensor output to an analog or digital input pin on the Arduino. Connect the LED and its resistor to a digital output pin.

- **Programming:** Write a simple program to monitor the sensor's input and control the LED. For example:

int sensorPin = A0; // Photoresistor connected to analog pin

int ledPin = 9;   // LED connected to digital pin

```
void setup() {

  pinMode(ledPin, OUTPUT);

  Serial.begin(9600);

}


void loop() {

  int sensorValue = analogRead(sensorPin);

  Serial.println(sensorValue);  // Monitor sensor value


  if (sensorValue < 500) {  // Adjust threshold based on your setup

    digitalWrite(ledPin, HIGH);  // Turn on LED

  } else {

    digitalWrite(ledPin, LOW);  // Turn off LED

  }

  delay(100);
```

}

This approach allows fine-tuning of thresholds and even the addition of multiple conditions (e.g., motion and light sensing combined).

Step 4: Testing and Troubleshooting

- Use a multimeter to verify the sensor's output. Check that the voltage changes as expected when the environmental condition (light or motion) changes.
- If the LED doesn't light up, check the connections to the transistor and the orientation of the LED.
- For PIR sensors, ensure the trigger delay is set correctly. Many PIR modules have adjustable delay and sensitivity knobs.

Applications and Extensions

Once you've built the basic system, consider expanding its functionality:

- **Multi-Sensor Integration:** Combine light and motion sensors to create a more sophisticated lighting system. For example, only light the LEDs if it's dark and motion is detected.
- **Energy Efficiency:** Add a timer or a dimming feature to conserve power.
- **Decorative Applications:** Use RGB LEDs and program different colors based on sensor inputs.

Creating a dynamic LED lighting system is more than just a fun project—it's a gateway to understanding how sensors, logic, and control work together in modern electronics. By completing this project, you'll gain valuable experience with condition-based control, sensor integration, and circuit design, setting the stage for even more advanced systems.

7.3 Building a Solar-Powered USB Charger

Harnessing solar energy to power devices is an excellent way to explore renewable energy systems while creating something practical. In this project, you'll design and build a solar-powered USB charger capable of charging smartphones or other USB devices. This project introduces solar panels, power conversion, battery integration, and efficiency optimization—all vital concepts for renewable energy applications.

Understanding the Basics: Solar Panels and Power Conversion

Solar panels convert sunlight into electrical energy using photovoltaic (PV) cells. The output of a small solar panel typically ranges from 5V to 18V, depending on its size and the amount of sunlight it receives. However, USB devices require a stable 5V supply, meaning you'll need a way to regulate the panel's variable output into a consistent voltage. In this project, a DC-DC buck converter will step down the panel's output to 5V. If you choose to include a battery for energy storage, additional circuitry will manage charging and discharging, ensuring the system operates safely and efficiently.

Components and Tools Needed

To build the solar-powered USB charger, you'll need the following:

- Solar Panel: A 6V to 12V solar panel (with a power output of at least 5W).
- Buck Converter: A DC-DC step-down converter module to regulate voltage to 5V.
- Rechargeable Battery: A lithium-ion or lithium-polymer battery for energy storage (e.g., 3.7V 18650 cell).
- Battery Management System (BMS): A circuit to protect the battery from overcharging and over-discharging.
- USB Output Module: A 5V USB-A or USB-C module for charging devices.

- Diode: A Schottky diode to prevent backflow of current from the battery to the solar panel.
- Tools: Soldering iron, wires, breadboard (optional for prototyping), and multimeter for testing.

Step 1: Connecting the Solar Panel

Start by testing your solar panel under direct sunlight. Use a multimeter to measure its open-circuit voltage and short-circuit current to ensure it matches the specifications. Connect the positive terminal of the panel to a Schottky diode. This prevents the battery from discharging back into the panel when it's not generating power, such as during the night or in low light.

Step 2: Regulating Voltage with a Buck Converter

The output of the solar panel is likely too high and variable for USB devices. A buck converter steps down this voltage to a stable 5V. Adjust the converter's output by turning its potentiometer (if adjustable) and verifying the output voltage with a multimeter. Connect the diode's output to the input of the buck converter, ensuring the positive and negative connections are correct. The output of the converter will later be connected to the USB module or battery.

Step 3: Adding a Battery for Storage

To allow charging even when the sun isn't shining, integrate a rechargeable battery into the system. Connect the battery to a BMS circuit, which ensures the battery operates safely by preventing overcharging, over-discharging, and excessive current draw. The solar panel (through the buck converter) will connect to the BMS input, while the BMS output will feed the USB module.

Step 4: USB Charging Circuit

Attach a USB output module to the system. This module takes the regulated 5V output and provides a USB-A or USB-C port for connecting devices. If you're not using a battery, connect the module directly to the buck converter's output. With a battery, the module connects to the BMS output. Ensure proper wiring to avoid polarity issues.

Step 5: Assembling the Circuit

Once all components are connected and verified individually, assemble them on a breadboard or solder them onto a prototyping board for a more permanent setup. Arrange the solar panel, buck converter, battery, BMS, and USB module so they're securely connected. Use a small enclosure to house the components if you want a more polished design.

Step 6: Testing and Troubleshooting

Place the solar panel in direct sunlight and connect a USB device to the output. Use a multimeter to monitor the voltage at different points in the circuit, ensuring the system delivers a steady 5V to the USB output. If the system doesn't work as expected, check for:

- Incorrect wiring or loose connections.
- Insufficient sunlight or a panel that doesn't meet the required power output.
- A misconfigured buck converter or a faulty BMS circuit.

Optimizing Efficiency

To improve the charger's performance, consider using higher-efficiency components, such as Maximum Power Point Tracking (MPPT) modules for solar panels. These devices ensure the panel operates at its optimal power output under varying sunlight conditions. Additionally, using high-quality batteries and minimizing voltage drops in the circuit will enhance overall efficiency.

Applications and Extensions

This project is a great starting point for exploring renewable energy. You can expand its functionality by:

- Adding an LCD or LED indicator to display battery charge levels.
- Using a larger solar panel and multiple batteries to charge higher-capacity devices or multiple devices simultaneously.
- Incorporating wireless charging for compatible devices.

A solar-powered USB charger is a practical and rewarding project that introduces you to renewable energy systems and basic power electronics. By the end of this project, you'll have gained hands-on experience with solar panels, power conversion, and battery management, equipping you with skills that can be applied to more complex renewable energy projects in the future.

7.4 Designing a Temperature Sensor with Microcontroller Integration

Designing a temperature sensor that interfaces with a microcontroller is a hands-on project that combines sensing, data processing, and display integration. In this project, you'll use a temperature sensor to measure the ambient temperature, process the data with a microcontroller like Arduino, and display the readings on an LCD or monitor. This project introduces sensor interfacing, basic programming, and debugging, helping you understand how microcontrollers interact with the physical world.

Project Overview

This system will use a temperature sensor, such as the LM35 or DHT11, to read temperature data. The sensor's output will be sent to a microcontroller, which will process the signal and display the temperature on an LCD or in the Arduino IDE serial monitor. The project introduces the core concepts of reading and interpreting sensor data and translating it into a human-readable format.

Components and Tools Needed

- **Microcontroller:** Arduino Uno or a compatible board.
- **Temperature Sensor:**
    - LM35 (analog sensor): Outputs a voltage proportional to the temperature.
    - DHT11 or DHT22 (digital sensor): Outputs temperature and humidity data in a digital format.
- **Display Module:** 16x2 LCD (optional: I2C adapter for simpler wiring) or a serial monitor via the Arduino IDE.
- **Resistors:** If needed for the sensor or pull-up configuration.
- **Breadboard and Jumper Wires:** For assembling the circuit.
- **Power Supply:** USB cable for the Arduino or an external power source.
- **Programming Software:** Arduino IDE for writing and uploading code.

Step 1: Understanding the Temperature Sensor

Choose the sensor based on your needs:

- **LM35:** Outputs an analog voltage directly proportional to the temperature. Each 10mV represents 1°C, making it easy to calculate temperature values.
- **DHT11/DHT22:** Digital sensors that provide both temperature and humidity data. Communication is handled via a single data pin, using a digital protocol.

Step 2: Wiring the Circuit

1. **LM35 Connection:**
    - Connect the sensor's VCC pin to 5V on the Arduino.
    - Connect the GND pin to ground.
    - Connect the output pin to an analog input pin on the Arduino (e.g., A0).
2. **DHT11/DHT22 Connection:**
    - Connect the VCC pin to 5V and GND to ground.

- Connect the data pin to a digital input pin (e.g., D2). Use a 10kΩ pull-up resistor between the data pin and VCC if required.
3. **LCD Display (Optional):**
    - For a 16x2 LCD, connect the RS, E, and data pins to the Arduino's digital pins, along with power and ground.
    - If using an I2C adapter, connect the SDA and SCL pins to the corresponding Arduino pins (A4 and A5 on an Uno).

Step 3: Writing the Code

For this project, the code will read data from the sensor, process it, and output the temperature to the display or serial monitor.

1. **LM35 Example Code:**

```
const int sensorPin = A0; // LM35 connected to A0

float temperature;


void setup() {

  Serial.begin(9600);

}


void loop() {

  int sensorValue = analogRead(sensorPin);

  temperature = sensorValue * (5.0 / 1023.0) * 100.0; // Convert to Celsius
```

```
  Serial.print("Temperature: ");

  Serial.print(temperature);

  Serial.println(" °C");

  delay(1000);

}
```

DHT11 Example Code:
Install the DHT library in the Arduino IDE (available via the Library Manager).

```
#include "DHT.h"


#define DHTPIN 2     // DHT sensor connected to pin 2

#define DHTTYPE DHT11 // DHT11 sensor


DHT dht(DHTPIN, DHTTYPE);


void setup() {

  Serial.begin(9600);

  dht.begin();

}
```

```
void loop() {

  float temperature = dht.readTemperature();

  if (isnan(temperature)) {

    Serial.println("Error reading temperature!");

    return;

  }

  Serial.print("Temperature: ");

  Serial.print(temperature);

  Serial.println(" °C");

  delay(1000);

}
```

7.5 Constructing a DC Motor Control System

Controlling the speed and direction of a DC motor is a fundamental project in electronics, showcasing the principles of motor control and circuit design. In this project, you'll build a circuit that uses Pulse Width Modulation (PWM) and an H-bridge motor driver to adjust motor speed and direction. This hands-on experience introduces motor control theory and its practical applications, laying the groundwork for more advanced projects like robotic arms, automated vehicles, or conveyor systems.

Project Overview

A DC motor operates by converting electrical energy into mechanical motion. To control its speed, we use PWM to vary the average voltage supplied to the motor. To control direction, an H-bridge circuit allows the current to flow in either direction through the motor, reversing its rotation. Combined, these techniques give full control over the motor's operation.

Components and Tools Needed

- **DC Motor:** A small hobby motor or any DC motor suitable for the project.
- **Motor Driver IC:** L298N or L293D (commonly used H-bridge motor driver modules).
- **Microcontroller:** Arduino Uno or similar board to generate the PWM signals.
- **Power Supply:** Battery pack or DC adapter (e.g., 6V–12V) to power the motor.
- **Diodes:** If building your own H-bridge circuit.
- **Push Buttons or Potentiometer:** Optional, for manual speed and direction control.
- **Breadboard and Jumper Wires:** For circuit assembly.
- **Multimeter:** To test and troubleshoot connections.

Step 1: Understanding PWM and H-Bridge Motor Drivers

PWM involves rapidly switching the motor's power supply on and off at a specific frequency. The ratio of the ON time to the total cycle time (duty cycle) determines the average voltage and, consequently, the motor's speed.

An H-bridge is a circuit that uses switches (transistors or MOSFETs) to allow current to flow in either direction through the motor, reversing its rotation. Prebuilt motor driver modules like the L298N simplify this process, integrating an H-bridge and other necessary components.

Step 2: Wiring the Circuit

1. **Connect the Motor Driver to the Arduino:**
    - Connect the motor terminals to the output pins of the motor driver module.

- Connect the motor driver's input pins (IN1, IN2 for direction control and ENA for PWM speed control) to the Arduino's digital output pins (e.g., D3, D4, D5).
- Provide power to the motor driver's VCC pin (6V–12V for the motor) and logic power to the 5V pin (from the Arduino). Ground the motor driver and Arduino together.

2. **Connect the Arduino to the Motor Driver:**
    - Digital pins on the Arduino control the direction by toggling IN1 and IN2 HIGH or LOW.
    - The ENA pin controls the speed using a PWM signal from the Arduino.
3. **Add a Potentiometer (Optional):**
    - Connect a potentiometer to an analog input pin on the Arduino. This will allow you to adjust the motor speed dynamically by varying the PWM duty cycle.

Step 3: Writing the Code

Use the Arduino IDE to write a program that generates PWM signals and controls motor direction. Below is a sample code:

```
const int enA = 9; // PWM pin for speed control

const int in1 = 7; // Motor direction control pin 1

const int in2 = 8; // Motor direction control pin 2


void setup() {

  pinMode(enA, OUTPUT);

  pinMode(in1, OUTPUT);

  pinMode(in2, OUTPUT);
```

}

```
void loop() {
  // Rotate motor forward at 50% speed
  digitalWrite(in1, HIGH);
  digitalWrite(in2, LOW);
  analogWrite(enA, 128); // 50% duty cycle (0-255)
  delay(2000);

  // Rotate motor backward at full speed
  digitalWrite(in1, LOW);
  digitalWrite(in2, HIGH);
  analogWrite(enA, 255); // 100% duty cycle
  delay(2000);

  // Stop the motor
  digitalWrite(in1, LOW);
  digitalWrite(in2, LOW);
  analogWrite(enA, 0);
```

    delay(2000);

}

Step 4: Testing the Circuit

Upload the code to the Arduino and power the circuit. The motor should rotate forward, backward, and stop according to the program. If it doesn't work as expected:

- Check the wiring, ensuring all connections are secure and correct.
- Use a multimeter to verify the voltage at the motor terminals during each operation.
- Verify the motor driver is receiving signals from the Arduino.

Step 5: Adding Advanced Features

Once the basic system is working, consider adding more features:

- **Speed Control via Potentiometer:** Replace the fixed PWM duty cycle in the code with values read from a potentiometer.
- **Direction Control via Buttons:** Use push buttons to toggle motor direction.
- **Automated Control:** Add sensors to control the motor's behavior based on environmental inputs (e.g., a line-following robot).

Applications of Motor Control Systems

Motor control is a cornerstone of many practical applications, including robotics, conveyor systems, and smart appliances. The techniques used in this project can be extended to control multiple motors, integrate feedback systems like encoders, or automate complex tasks.

Constructing a DC motor control system introduces you to PWM and H-bridge circuits, providing valuable experience in combining theory with practical implementation. By mastering these skills, you'll be equipped to tackle more advanced projects involving motion control and automation.

7.6 Final Tips for Taking Your Projects Further

Completing a hands-on project is an exciting achievement, but it's only the beginning of your journey in electrical engineering. The real value comes from refining, customizing, and scaling your work. By building on what you've created, exploring new techniques, and learning from every challenge, you can take your projects to the next level. Here are some tips to guide you.

1. Customize and Experiment

Every project you build is an opportunity to experiment. Think about how you can enhance functionality or add unique features. For example, in your LED lighting system, you could incorporate RGB LEDs to create dynamic color effects. For the solar charger, adding an LCD to display battery status or charging current could make it more user-friendly. Experimentation is where creativity meets problem-solving and can lead to truly innovative designs.

Customization also helps you learn how different components interact. Try swapping out a standard DC motor for a stepper motor in your control system or using a different type of sensor in your temperature monitoring project. These small changes teach you how to adapt to varying requirements and constraints.

2. Source Components Effectively

Finding the right components at reasonable prices is a crucial skill. While local electronics stores are convenient for small purchases, online platforms like Digi-Key, Mouser, and SparkFun often offer a broader selection and better prices. For hobbyists, marketplaces like Amazon or AliExpress can be great for acquiring modules, sensors, and basic kits.

When sourcing, pay attention to component specifications. Ensure they meet your project's requirements for voltage, current, and compatibility. Reading reviews or joining forums like Reddit's r/Electronics can help you identify reliable products and vendors.

3. Learn PCB Design

As your projects become more complex, breadboards and jumper wires may not suffice. Learning to design and manufacture Printed Circuit Boards (PCBs) can elevate your work. Tools like KiCad, Eagle, and Altium Designer are popular for creating custom PCBs. A well-designed PCB not only organizes your components but also improves reliability and reduces errors.

Once you've designed a PCB, services like JLCPCB or PCBWay can fabricate your boards at an affordable cost. Seeing your design come to life in a professionally produced circuit board is a rewarding experience that adds polish to your projects.

4. Embrace Failure as a Learning Opportunity

Not every project will work perfectly on the first try. Circuits may fail, sensors might not behave as expected, or software bugs can disrupt functionality. Instead of getting discouraged, view these moments as valuable lessons. Document what went wrong, investigate the cause, and iterate on your design. Many groundbreaking ideas emerge from trial and error.

5. Document Your Work

Keeping detailed records of your projects is essential. Take notes on your design process, challenges, and solutions. Draw circuit diagrams, save your code, and take photos of your setups. Documentation not only helps you revisit and refine your work but also makes it easier to share your knowledge with others. Platforms like GitHub are excellent for organizing and sharing your code and schematics with the engineering community.

6. Seek Feedback and Collaborate

Electrical engineering is as much about community as it is about individual effort. Share your projects on platforms like Hackaday.io, Reddit, or specialized forums. Seeking feedback can provide valuable insights, while collaboration with others might lead to new ideas or improvements.

Mentors can be particularly valuable as you grow. Experienced engineers can guide you through complex challenges, recommend resources, or offer advice based on their own experiences. Don't hesitate to reach out to local engineering clubs or online communities to find mentors willing to help.

7. Scale Your Projects

Once you've mastered small projects, think bigger. Could your solar charger power multiple devices? Can your motor control system integrate with a larger robot? Scaling requires you to consider factors like power management, component selection, and system reliability, which are critical for real-world applications.

Keep Building, Keep Learning

The beauty of electrical engineering lies in its endless possibilities. Every project you complete builds your skills and confidence, preparing you for even greater challenges. By experimenting, documenting, and collaborating, you can transform small projects into impactful innovations. Most importantly, enjoy the process—each circuit you build brings you closer to mastering the art of engineering.

Chapter 8: The Future of Electrical Engineering

8.1 Emerging Technologies in Electrical Engineering

Electrical engineering is a dynamic field, continually evolving to address modern challenges and opportunities. The rapid advancement of technology has given rise to groundbreaking innovations that are redefining what is possible in power systems, electronics, and computing. From quantum technologies to nanotechnology, electrical engineers are at the forefront of shaping the future. This section explores some of the most exciting emerging technologies and their intersections with other fields, highlighting the profound impact of electrical engineering in diverse areas.

Quantum Computing: The Next Frontier in Processing Power

Quantum computing represents a paradigm shift in computational power and capability. Unlike traditional computers, which process information as binary bits (0s and 1s), quantum computers use quantum bits or qubits, capable of existing in multiple states simultaneously thanks to quantum superposition.

This technology has transformative potential for electrical engineering, particularly in optimizing power grids, designing advanced materials, and enhancing signal processing. For instance, quantum algorithms can solve complex problems in seconds that would take classical computers millennia to compute. Electrical engineers play a crucial role in designing the hardware for quantum systems, including quantum gates, error correction circuits, and cryogenic systems for maintaining the delicate quantum states.

Applications:

- Optimizing large-scale power systems for renewable energy integration.

- Developing quantum sensors for ultra-precise measurements in healthcare and aerospace.
- Revolutionizing cryptography, ensuring secure communications in an increasingly connected world.

Artificial Intelligence in Power Systems

Artificial Intelligence (AI) is rapidly becoming an integral part of electrical engineering, particularly in the management and optimization of power systems. AI algorithms can analyze vast amounts of data from sensors and smart grids, enabling real-time decision-making and predictive maintenance.

For example, machine learning models are used to forecast energy demand, allowing utilities to adjust generation and distribution proactively. Similarly, AI-powered systems can detect faults in power lines or transformers before they escalate, minimizing downtime and repair costs.

In renewable energy, AI helps optimize the output of solar panels and wind turbines by analyzing environmental conditions and dynamically adjusting operations. It also plays a key role in developing energy-efficient appliances and industrial systems.

Applications:

- Smart grids that self-heal during outages and adapt to changing demands.
- Intelligent home energy management systems that reduce waste.
- Enhanced battery management in electric vehicles (EVs) for longer ranges and faster charging.

Nanotechnology in Electronics

Nanotechnology is revolutionizing electronics by enabling the creation of devices at the atomic and molecular levels. Nanoelectronics, which deals with components at the nanometer scale, is paving the way for faster, smaller, and more energy-efficient devices.

One breakthrough example is the development of transistors made from graphene, a single layer of carbon atoms with extraordinary electrical properties. These transistors have the potential to outperform traditional silicon-based components, leading to faster processors and lower power consumption.

Nanotechnology also plays a vital role in advanced sensors, such as nanoscale pressure or chemical sensors, which are used in healthcare and environmental monitoring. Furthermore, nanoscale fabrication techniques are crucial in building quantum dots and LEDs for ultra-high-definition displays and efficient lighting systems.

Applications:

- Ultra-fast processors for next-generation computing.
- Flexible and wearable electronics, including smart fabrics and medical devices.
- High-efficiency solar panels and energy storage solutions.

Intersections with Biotechnology

Electrical engineering intersects with biotechnology in developing life-changing medical devices and systems. From advanced imaging equipment like MRI and CT scanners to implantable devices such as pacemakers, engineers are enabling better healthcare outcomes through innovation.

Emerging areas include neural engineering, where electronics interface directly with the human brain to restore lost functions, such as in prosthetic limbs controlled by thought. Wearable devices like fitness trackers and continuous glucose monitors are also becoming more sophisticated, offering real-time health monitoring powered by microcontrollers and sensors.

Applications:

- Brain-computer interfaces for patients with mobility impairments.
- Biosensors for detecting diseases early and monitoring chronic conditions.
- Lab-on-a-chip devices for rapid diagnostic testing.

Advancements in Renewable Energy

Renewable energy systems are evolving rapidly, thanks to innovations in electrical engineering. Advances in power electronics, such as highly efficient inverters and converters, have made solar and wind energy systems more reliable and cost-effective.

Energy storage is also benefiting from breakthroughs in battery technology, such as solid-state batteries that offer greater energy density and safety. Smart grid technologies, as discussed in **4.5 Basic Overview of Renewable Energy Systems**, enable better integration of renewable energy into existing infrastructure, reducing reliance on fossil fuels.

Applications:

- Floating solar farms for efficient space utilization.
- Wind turbines with predictive maintenance to reduce downtime.
- Microgrids for remote communities, offering localized and resilient energy solutions.

Exploring Space with Electrical Engineering

The role of electrical engineering in space exploration is profound. From powering spacecraft to designing communication systems, electrical engineers make interplanetary exploration possible. Solar arrays provide energy for satellites and rovers, while advanced electronics handle navigation, communication, and scientific instruments.

Emerging technologies like ion propulsion and nuclear-powered generators are extending the range and duration of space missions. Electrical engineers are also working on power systems for habitats on the Moon or Mars, ensuring they are sustainable and efficient.

Applications:

- Lightweight, high-efficiency solar panels for deep space missions.
- Radiation-hardened electronics for harsh space environments.
- Power and communication systems for planetary exploration vehicles.

8.2 The Role of Sustainability and Renewable Energy

Sustainability has become a critical focus in electrical engineering, driven by the urgent need to address climate change and reduce environmental impact. Engineers today are tasked with designing systems and technologies that balance efficiency, cost, and ecological responsibility. Renewable energy systems, advancements in energy storage, and carbon-neutral innovations are transforming the field, making sustainable practices not just a priority but a necessity.

Sustainable practices in engineering begin with mindful resource management. This includes minimizing energy consumption, using recyclable or biodegradable materials, and designing systems that are both efficient and durable. Electrical engineers contribute by creating technologies that optimize energy usage, such as energy-efficient appliances, smart lighting, and industrial automation systems. Moreover, sustainable engineering extends to the entire lifecycle of a product, from manufacturing to disposal, ensuring minimal waste and environmental impact.

Renewable energy systems are central to sustainability efforts, offering alternatives to fossil fuels. Technologies such as solar and wind power have become more accessible due to advances in electrical engineering. As mentioned in **4.5 Basic Overview of Renewable Energy Systems**, power electronics like inverters and converters play a crucial role in integrating renewable sources into grids. Engineers are now focusing on enhancing the efficiency and reliability of these systems, making renewable energy a viable option for global energy needs.

Energy storage is another area where sustainability is driving innovation. Traditional lithium-ion batteries are giving way to advanced technologies like solid-state batteries, which promise greater energy density, faster charging, and improved safety. Solid-state batteries replace the liquid electrolyte with a solid one, reducing the risk of leakage and fire while extending battery life. These advancements are crucial for renewable energy storage and electric vehicles, where efficient, long-lasting batteries are essential for widespread adoption.

Smart grids are revolutionizing how electricity is managed and distributed. These intelligent systems use sensors, data analytics, and automation to optimize energy flow, reduce losses, and integrate renewable sources effectively. Smart grids can adapt to real-time demand and supply changes, ensuring stability even with intermittent energy sources like solar and wind. For example, during peak sunlight hours, excess solar power can be stored in batteries or redistributed to areas with higher demand, reducing reliance on traditional power plants.

Carbon-neutral technologies are also making significant strides. Engineers are developing solutions that offset or eliminate carbon emissions, such as carbon capture and storage (CCS) systems that trap CO_2 from industrial processes. In addition, the adoption of green hydrogen, produced using renewable energy, is creating new pathways for decarbonizing industries like steel production and long-haul transportation. Electrical engineers are critical in designing the power systems and infrastructure needed for these technologies to scale.

Sustainability in electrical engineering isn't just about addressing current challenges; it's about future-proofing systems to meet long-term environmental goals. For example, microgrids, which operate independently or in conjunction with the main grid, are being deployed in remote or disaster-prone areas. These systems provide reliable energy while incorporating renewable sources, reducing dependence on fossil fuels and increasing resilience.

By embracing sustainability, engineers can lead the transition to a cleaner, more equitable energy future. Through innovative designs and thoughtful resource management, electrical engineers have the tools to reduce environmental impact while meeting the growing demand for energy. The integration of renewable energy, advanced storage technologies, and smart systems represents a monumental shift in how energy is generated, stored, and used.

Sustainable practices in electrical engineering not only address environmental concerns but also create economic and social benefits. Lower energy costs, improved energy access, and reduced greenhouse gas emissions are just a few of the advantages that come from integrating sustainability into engineering. By prioritizing these goals, the field can continue to innovate while safeguarding the planet for future generations.

8.3 Challenges Facing the Industry

As electrical engineering continues to drive technological progress, the industry faces several challenges that require careful consideration and innovative solutions. From ensuring the security of smart devices to addressing ethical dilemmas in automation and bridging the gap in global energy accessibility, engineers are tasked with navigating complex issues that have far-reaching implications. Tackling these challenges responsibly will be key to ensuring sustainable and equitable progress in the field.

One of the most pressing challenges is **cybersecurity in smart devices and systems**. The rise of the Internet of Things (IoT) has connected billions of devices, from household appliances to industrial control systems, creating a vast network of interconnected technology. While these innovations improve convenience and efficiency, they also introduce vulnerabilities. Cyberattacks targeting power grids, smart homes, and medical devices could have catastrophic consequences.

To address these risks, electrical engineers must design systems with robust security measures. This includes encrypting communications between devices, implementing multi-factor authentication, and using hardware-based security features. Developing protocols that ensure data integrity and protect against unauthorized access is essential. Collaboration with cybersecurity experts and ongoing education in secure design practices can help engineers stay ahead of evolving threats.

Another critical issue is the **ethical considerations surrounding automation and artificial intelligence (AI)**. Automation has the potential to revolutionize industries, increasing productivity and reducing human error. However, it also raises ethical concerns, such as job displacement, decision-making accountability, and the potential for biased algorithms. For example, in autonomous vehicles, ethical dilemmas arise when programming decision-making systems to handle life-and-death situations.

Engineers must approach automation projects with a focus on transparency and fairness. Ethical frameworks, such as IEEE's Global Initiative on Ethics of Autonomous and Intelligent Systems, provide guidelines for responsible development. Additionally, engineers should work closely with interdisciplinary teams, including ethicists and policymakers, to ensure that automation solutions align with societal values and promote equity.

Global energy accessibility remains a significant challenge, particularly in developing regions where millions lack reliable electricity. As discussed in **8.2 The Role of Sustainability and Renewable Energy**, renewable energy systems like microgrids offer promising solutions. However, implementing these systems in remote or under-resourced areas comes with logistical, financial, and technical hurdles.

Engineers must focus on designing affordable, scalable, and resilient energy systems that can be deployed in diverse environments. This includes using locally available materials, simplifying maintenance requirements, and integrating off-grid solutions like solar-powered microgrids. Partnerships between governments, non-profits, and private companies can provide the funding and resources needed to bring these projects to fruition.

Resource scarcity and environmental impact further complicate the industry's efforts to innovate sustainably. The demand for rare-earth metals used in batteries, motors, and electronics continues to grow, leading to supply chain challenges and environmental concerns associated with mining. Developing alternative materials, improving recycling processes, and designing systems that minimize material usage are crucial for mitigating these impacts.

The increasing complexity of electrical systems also presents **technical challenges in integration and standardization**. For instance, integrating renewable energy sources into existing power grids requires careful management to maintain stability and reliability. Similarly, the lack of standardized protocols for IoT devices can hinder interoperability and create inefficiencies. Engineers must prioritize collaboration and adopt flexible, forward-compatible designs to address these issues.

Despite these challenges, the electrical engineering industry is uniquely positioned to lead the way with innovative and responsible solutions. Engineers can leverage advancements in machine learning to enhance cybersecurity, apply ethical principles to ensure fair automation practices, and use renewable technologies to address energy inequality. Furthermore, fostering a culture of collaboration and lifelong learning will empower professionals to adapt to evolving demands and drive positive change.

The path to overcoming these challenges lies in the commitment of engineers to innovation, ethical responsibility, and adaptability. Tackling issues like cybersecurity, automation ethics, and energy accessibility requires a proactive approach that balances technological advancement with the needs of society and the environment. Engineers are uniquely equipped to craft solutions that address these pressing problems, transforming obstacles into opportunities for progress. By fostering collaboration, embracing lifelong learning, and prioritizing sustainability, the industry can pave the way toward a secure, equitable, and sustainable future.

8.4 Opportunities for the Next Generation of Engineers

The next generation of electrical engineers is entering the field at a pivotal moment in history. Rapid technological advancements, growing sustainability demands, and the expansion of interdisciplinary innovation have created unprecedented opportunities. Aspiring engineers who can align their skills with these emerging trends will find themselves well-positioned to make a meaningful impact while building rewarding careers.

1. Startups and Entrepreneurial Ventures

The rise of technology-focused startups has opened doors for engineers with innovative ideas and a drive for entrepreneurship. Startups thrive on disruption, often targeting niche markets with solutions that traditional industries overlook. For electrical engineers, this could mean developing IoT devices that address everyday challenges, creating custom renewable energy solutions for small businesses, or launching a company focused on advanced robotics.

Working in or founding a startup allows engineers to experience multiple roles—from design and prototyping to business strategy and customer interaction. It's an excellent environment for rapid skill growth and fostering creativity. Aspiring engineers should consider developing a strong foundation in business basics, such as product management and funding strategies, to complement their technical expertise.

2. Opportunities in Interdisciplinary Fields

Electrical engineering increasingly overlaps with fields like biotechnology, data science, and environmental engineering. Engineers who expand their knowledge into these areas can access roles that were once out of reach for traditional specialists. For example, electrical engineers can contribute to the development of wearable medical devices by combining expertise in embedded systems with a basic understanding of physiology. Similarly, collaboration with environmental scientists can lead to innovations in sustainable energy systems or smart agriculture.

Emerging fields such as quantum computing and neuromorphic engineering also offer exciting opportunities. Engineers working in these areas need a solid understanding of electrical principles while exploring adjacent disciplines like physics or neuroscience. Developing interdisciplinary skills can broaden career prospects and lead to pioneering innovations.

3. Research-Driven Roles

For those drawn to exploration and discovery, research presents a compelling career path. Electrical engineers in research roles work on solving fundamental problems and pushing the boundaries of what's possible. Whether in academia, government agencies, or corporate R&D labs, researchers are instrumental in creating new materials, devices, and systems.

Research-driven roles offer the chance to delve deeply into topics like advanced semiconductors, smart grid optimization, or AI integration in electronics. Engineers in this space often work on long-term projects with global implications, such as improving battery technology to support widespread adoption of renewable energy. Aspiring researchers should pursue advanced degrees and stay connected with academic conferences and journals to keep pace with developments in their chosen area of focus.

4. Preparing for the Future

To align with future industry demands, aspiring engineers must embrace a mindset of continuous learning and adaptability. As industries evolve, so do the skills required to succeed. Keeping up with emerging technologies, mastering industry-standard tools, and pursuing certifications are all effective ways to stay relevant. For example, familiarity with programming languages like Python or MATLAB, and tools such as SPICE or Altium Designer, can provide a competitive edge.

Networking is another critical element of career development. Joining professional organizations such as IEEE or attending hackathons and engineering meetups can help aspiring engineers connect with peers, mentors, and potential collaborators. Additionally, internships and co-op programs offer valuable hands-on experience and insight into different industries, helping young engineers refine their career goals.

5. Balancing Impact and Passion

The future of electrical engineering is about more than just technology—it's about creating meaningful change. Aspiring engineers should seek roles that align with their passions while addressing pressing global challenges. For some, this might mean working on renewable energy solutions to combat climate change, while others may find fulfillment in developing medical devices that improve lives.

Carefully considering personal interests and the impact of one's work can lead to a fulfilling career. Opportunities abound for engineers to apply their skills in ways that contribute to a better world, whether through technological innovation, sustainable practices, or collaboration across disciplines.

A Future of Endless Possibilities

The opportunities awaiting the next generation of electrical engineers are vast and varied. By aligning their skills with industry demands, embracing interdisciplinary collaboration, and pursuing meaningful work, aspiring engineers can thrive in a world increasingly shaped by technology. The future is a blank canvas, and engineers hold the tools to create solutions that inspire, transform, and endure.

8.5 Inspiring the Engineer Within You

Electrical engineering is more than a career—it's a journey of curiosity, problem-solving, and innovation. At its core, it is about shaping the future, solving pressing challenges, and improving lives through technology. As you continue your exploration of this exciting field, remember that the most successful engineers are those who approach their work with passion, creativity, and a relentless drive to learn.

The Power of Curiosity

Every great innovation begins with a simple question: "What if?" Electrical engineering is a field that rewards curiosity. It invites you to experiment, explore new ideas, and challenge the boundaries of what's possible. Whether you're designing a smart lighting system, optimizing renewable energy solutions, or contributing to cutting-edge research, your ability to ask the right questions will guide your path to discovery.

Curiosity drives you to dig deeper into problems, uncover hidden solutions, and connect seemingly unrelated concepts in novel ways. It's the spark that leads to breakthroughs and makes every project an opportunity to learn something new.

The Importance of Perseverance

Engineering is not without its challenges. Circuits fail, prototypes break, and solutions sometimes seem elusive. These moments can be frustrating, but they are also where growth happens. Perseverance is what separates a good engineer from a great one.

When faced with setbacks, remember that every failure is a step closer to success. Each obstacle teaches you valuable lessons about design, testing, and problem-solving. Celebrate small victories, stay patient, and never lose sight of the bigger picture. Great engineers are those who turn challenges into opportunities, approaching every problem with resilience and determination.

The Role of Creativity

At its heart, engineering is a creative endeavor. It's about finding new ways to solve old problems and imagining solutions to challenges that don't yet exist. Creativity allows you to think outside the box, combining technical expertise with innovative ideas to create systems and technologies that transform lives.

As an electrical engineer, your creativity could lead to groundbreaking advancements in renewable energy, smarter healthcare devices, or more efficient transportation systems. Embrace the freedom to experiment and innovate, knowing that your ideas have the power to make a difference.

Shaping the World

The impact of electrical engineers is everywhere. From the devices we use daily to the power grids that sustain our cities, electrical engineering shapes modern life. Your work has the potential to solve global challenges, from reducing carbon emissions to improving access to clean energy and healthcare.

As the world evolves, so too will the role of electrical engineers. You'll be at the forefront of addressing society's most pressing needs, using technology to drive progress and create a more sustainable, equitable future. The skills you've developed and the projects you've tackled are just the beginning of your contribution to this ongoing transformation.

A Call to Action

Now is your time to make your mark. Stay curious, embrace challenges, and let your creativity flourish. Seek out opportunities to learn and grow, collaborate with others, and pursue projects that ignite your passion. The possibilities are endless, and your potential is limitless.

Electrical engineering isn't just about solving problems—it's about creating a better world. As you move forward in your journey, carry with you the knowledge, skills, and inspiration to shape the future. The engineer within you is ready to rise to the challenge. Go forth and build something extraordinary.

Conclusion

As we reach the end of this journey through the world of electrical engineering, it's clear that this field is as vast as it is transformative. From mastering the basics of electricity to exploring advanced concepts like renewable energy and embedded systems, you've taken significant steps toward understanding how technology powers our world. Each chapter of this book has been designed to provide you with not just knowledge, but also the tools and inspiration to apply that knowledge in meaningful ways.

Electrical engineering is a discipline that thrives on curiosity, innovation, and perseverance. Whether you're building your first circuit, troubleshooting a complex system, or dreaming up groundbreaking solutions, the journey is always about learning and growth. The skills and concepts you've gained here are just the foundation—your creativity and determination will determine how far you can go.

As you continue to explore this exciting field, remember that the work of an engineer extends beyond technical challenges. It's about solving problems that matter, shaping technologies that improve lives, and contributing to a more sustainable and connected world. The future of electrical engineering is in your hands, and the possibilities are endless.

So, keep asking questions, stay resilient in the face of setbacks, and let your passion guide you. Whether you're designing the next breakthrough in electronics or finding innovative ways to power a community, your work will leave a lasting impact. This is just the beginning—your journey as an engineer is full of promise and potential. Go out there and make it count!

Bonus

Download your bonus here! Scan the QR code or open the link below

https://bit.ly/electricalengineeringbonus

Made in United States
Troutdale, OR
01/11/2025